Fourth edition

GEOGRAPHICAL, HISTORICAL AND POLITICAL MAPS OF CALIFORNIA

PATTERNS ON THE LAND

ROBERT W. DURRENBERGER

National Press Books, Palo Alto, California

SOURCES OF PHOTOGRAPHS:
C and H Sugar Corporation — 81; Cartwright Aerial Surveys — v; California Division of Highways — 14, 45, 47, 52; California Division of Mines — 39, 40; E. H. Grensted — 63; Hi Worth Pictures — 32; Library of Congress — 59, 71; Los Angeles County Museum — 23, 32, 42; Los Angeles Department of Airports — 86; Pacific Air Industries — 85; Redwood Empire Association — 84; San Diego Port Authority — 85; Southern Pacific Railroad — 88; U.S. Bureau of Reclamation — 66, 67, 76; U.S. Forest Service — 18; U.S. Soil Conservation Service — 20; University of California Extension Service — 26, 27, 28.

SOURCES OF MAPS, CHARTS AND DIAGRAMS:
Association of American Railroads — 88; California Department of Finance — 50, 51; California Department of Water Resources — 12, 79; Colorado River Association — 78; Los Angeles County Museum — 24, 32; National Automobile Club — 92 to 99; Erwin Raisz — 2 to 5; Rodney Steiner — 9, 77; U.S. Geological Survey — 8; *Weatherwise* — 11.

SOURCES OF DATA:
American Forest Products Industries — 68, 69; Bancroft, *History of California* — 26 to 39; California County Agricultural Commissioners — 62 to 67; California, Crop and Livestock Reporting Service — 62 to 67; California Department of Finance — 48 to 53, 58; California Department of Fish and Game — 70, 71; California Department of Industrial Relations — 82, 83; California Department of Water Resources — 76 to 79; California Division of Beaches and Parks — 80; California Division of Forestry — 68, 69; California Divisions of Mines and Geology — 6, 7, 41, 72, 73; *California Statistical Abstract* — 8; Caughey, *California* — 26 to 45; Driver and Massey, *North American Indians* — 24; Federal Aviation Authority — 87; Langer, *An Encyclopedia of World History* — 26 to 45; Paullin, *Atlas of Historical Geography* — 26 to 45; Rolle, *California, a History* — 26 to 45; Sale and Karn, *American Expansion* — 26 to 45; Sverdrup, *Oceanography for Meteorologists* — 22; U.S. Bureau of the Census — 48 to 53, 60, 63; U.S. Bureau of Commercial Fisheries — 70, 71; U.S. Corps of Engineers — 84, 85; U.S. Forest Service — 68, 69; U.S. Bureau of Indian Affairs — 25; U.S. Bureau of Land Management — 36, 37, 60, 61; U.S. Bureau of Mines — 72 to 75; U.S. Geological Survey — 8, 76 to 79; U.S. Soil Conservation Service — 20; U.S. Weather Bureau — 10 to 15; Western Oil and Gas Association — 74, 75.

International Standard Book No. 0-87484-207-7

Library of Congress Catalog Card No. MAP67-940

PREFACE TO THE FOURTH EDITION

California lends itself vividly to cartographical interpretation. Elements of the natural environment, such as its landforms, climates and soils are extraordinarily varied. Mt. Whitney, the highest mountain in the continental United States, is in California, as is Death Valley, the lowest place in the Americas. The cool and foggy coastal areas stand in marked contrast to the hot interior deserts and the cold and snowy areas in the mountains. There are regions of tall trees, large areas of grasses and shrubs, and some places so high or dry that only the hardiest small plants can grow.

Upon this varied natural landscape, men of different cultural backgrounds have left their imprint. Evidence of our heritage from the past is found in the place names of Indian and Spanish origin, the property boundaries pre-dating the rectangular survey, the old New England and Midwestern houses gracing some of our cities and the myriad of other cultural features which may be readily observed. The modern landscape presents scenes which range from those associated with one of the largest urban complexes found in the United States to those associated with the relatively untouched terrain in the High Sierra Country.

This diversity of physical and cultural background has created in the State of California an environment noteworthy for charm and interest unsurpassed in any part of the world. It seems appropriate that in an area with such a generous endowment of exciting vistas and tremendous works of man someone should attempt to capture and portray on maps as much of the total picture of the state as is possible. PATTERNS ON THE LAND represents an effort to do this. Still more, it represents an attempt to present to people everywhere a reference work which may be useful in understanding matters pertaining to California. In the pages which follow we have attempted to give the reader an accurate impression of the principal features of the natural environment; we have traced the historical development of the state; and we have portrayed the fundamental economic patterns.

We are indebted to many individuals, companies and governmental agencies for their contributions to the atlas. Most of them are listed on the opposite page. In addition, we have welcomed the suggestions and criticisms of our friends and colleagues who teach courses in California history and geography. Some of these suggestions have been incorporated into this atlas.

The production of the present edition would not have been possible without the diligent efforts of those individuals associated with the publication of the first three editions of the atlas. In particular, I would like to acknowledge the significant contributions of Dr. William Byron, John Kimura, Frank Burr and Harold Schwarm in the preparation of the earlier editions of the atlas.

ROBERT W. DURRENBERGER

CONTENTS

INTRODUCTION

There are patterns on the land in California, patterns which become discernible when the state is viewed as a whole, or when segments of the state are examined in detail. The individual must free himself from his immediate surroundings in order to see them. From the top of a high mountain or from the window of an airplane the apparent unsystematic arrangement of earth features begins to take on new meaning. Highways and rivers become lines on the land; houses agglomerate into settlements; and farms become agricultural regions.

The mosaic of natural and man-made features which comprise the California landscape is graphically portrayed in PATTERNS ON THE LAND. On occasion the reader views the state as it might be interpreted from a platform located some distance in space. At other times he is moved so close to the earth that only a small portion of California can be seen. Some elements have been mapped which cannot be seen at all, either because they occurred in the past, as in the case of pioneer movements, or because they reflect the statistical summary of a relatively long period of time, e.g., maps of average annual temperature and precipitation. In some cases the distributions have been superimposed on a relief base; in others they have not. Whatever base, scale and degree of generalization have been incorporated into individual maps, the atlas has been structured so that within any given series the reader may correlate data from two or more maps.

The atlas is organized to give the reader a systematic view of California's many landscape variations. The main body of the atlas is divided into four parts. Part I contains maps portraying various aspects of the natural environment. Photographs and textual material help the reader to understand the landforms, climate, and vegetation and soils of California. Part II presents the story of the exploration and settlement of California through a series of maps showing California as it was at different periods in the past, and textual material, graphs and photographs to illustrate better the movement of people into our state. Parts III, IV, and V represent an examination of man and his works in modern California. The Appendix contains sectional maps of California, and tables which amplify the data presented in the form of maps and charts.

The construction of freeways, homes, and businesses is changing the pattern of land use on the fringes of our major metropolitan areas.

To Californians, the need for an atlas such as this is apparent. In the past decade several million people have come into the state. Most of the new residents are transplanted Midwesterners, Easterners and Southerners—people who have lived in an environment entirely different from that found in California. The average citizen undoubtedly does not fully comprehend many of the fundamental facts which relate to the natural environment and its use. Such knowledge is a prerequisite to intelligent citizenship. It is essential if one is to cope with the problems associated with the determination of public policy. PATTERNS ON THE LAND is an attempt to make basic information available in a form which may be easily comprehended and quickly assimilated by our citizens. We hope that it will contribute to a better understanding of California.

A View southwest of Sacramento illustrating patterns of agricultural and urban land use; canal leading to the ocean extends south from the Sacramento Harbor.

Oil refineries and storage tanks occupy extensive areas in southwestern Los Angeles County.

2 LANDFORMS

I DEVONIAN PERIOD

II PERMIAN PERIOD

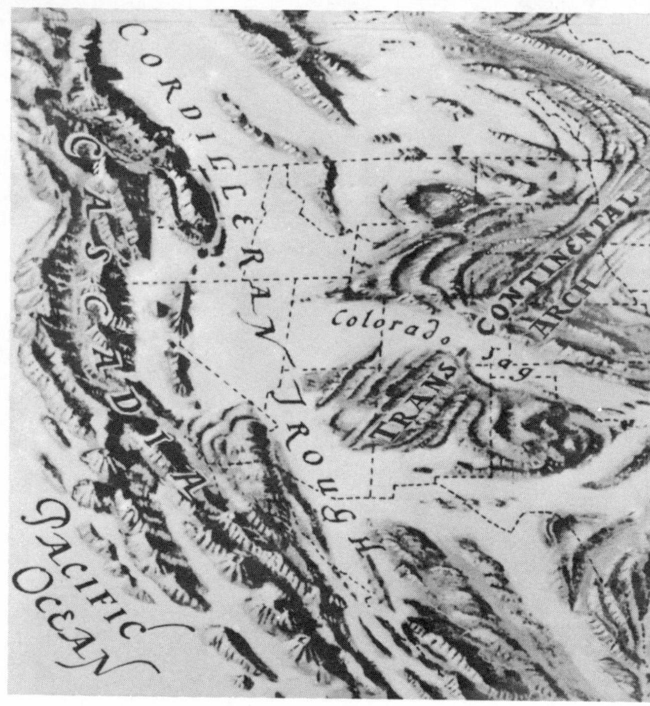

III JURASSIC PERIOD

IV TERTIARY PERIOD

Source: Erwin Raisz in Stovall and Brown's *The Principles of Historical Geology*.

EVOLUTION OF THE CALIFORNIA LANDSCAPE

The drawings on this page and on the opposite page illustrate the story of the evolution of the California landscape. Prior to 130,000,000 years ago a shallow arm of the ocean covered this area and successive layers of sedimentary rocks were deposited. A land mass known as Cascadia lay off to the west of present-day California while the mountainous mainland of North America was located to the east.

Then, portions of this giant basin were folded and faulted and molten rock material was forced upward from the interior of the earth; the ancestral Sierra Nevada was created. These mountains were subjected to repeated cycles of uplift and erosion interspersed with periods of volcanic activity until about 60,000,000 years ago when the ranges existed as low, undulating mountains with the ocean at their western edge. Then, further upheaval lifted the Sierra-Cascade System along an axis approximating its present position.

This upheaval was accompanied by faulting and volcanic action which covered the slopes and river valleys of the northern one-third of the Sierra Nevada with lava and produced lava flows and low volcanic domes in the Cascades and the Modoc Plateau.

Cascadia still lay off to the west, and coastal California was a zone of shallow basins and islands. In this zone there followed periods of mild folding and deposition in which parts of the basins were elevated and became islands while other parts submerged to become basins of deposition. Between 20,000,000 and 25,000,000 years ago pronounced earth movements established the general structural framework of the Coast Ranges. Gradually the mountains created by this activity were eroded, and a series of islands and shallow basins again occupied the zone. A period of relatively stable conditions ensued which lasted until about a million years ago when the entire Coast Range region was uplifted, folded and faulted and the sea receded from the Central Valley. At the same time renewed earth movements and volcanic activity in other parts of California modified the older landscape.

Water in the form of rain and snow fell on the land and collected in streams to wear down the hills and mountains. Periods when the agents of erosion were dominant were followed by periods

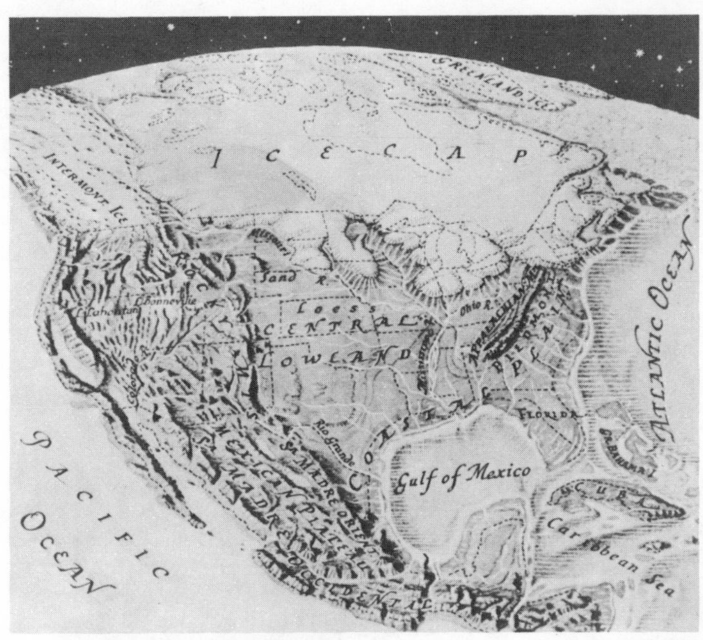

V QUATERNARY PERIOD

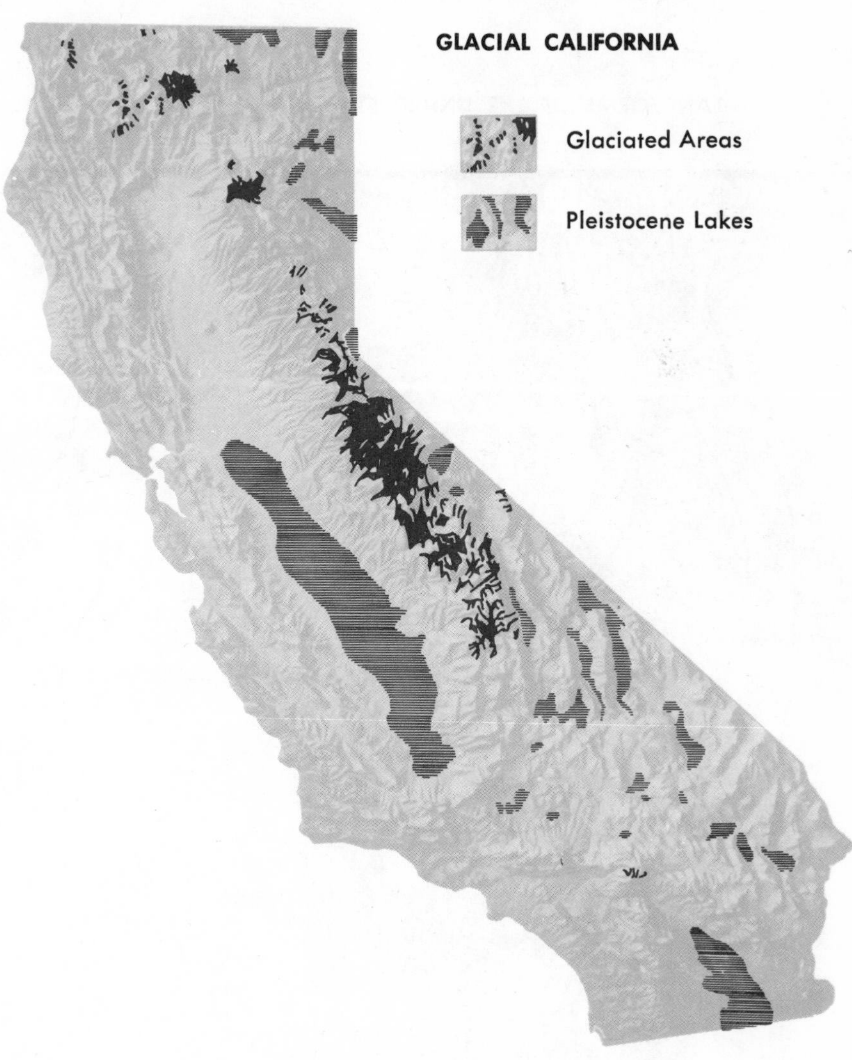

Volcanic activity has helped to create some of the landforms of California. Mt. Lassen is one of the best known peaks in the Cascade Range.

of active mountain building; the landscape was constantly changing.

Then, about the time that the Coast Ranges were taking final form one million years ago, snow and ice began to accumulate in the higher mountains of California and other parts of the West; continental glaciers formed over Canada and northern Europe. During the Ice Age glaciers extended out of the high mountain regions down the river valleys toward the lowlands. As they moved, the glaciers removed rock material from the mountain slopes and carried it down the valleys.

Over 10,000 years ago the climate became warmer and dryer and the glaciers melted. The rock material which they had been carrying was dropped. In many places in the High Sierra these glacial moraines have helped to create beautiful lakes. The water from the melting ice flowed out of the high mountain areas into the Central Valley and the desert basins and created large lakes in such places as Death Valley. Gradually, conditions changed and the present landscape emerged.

LANDFORMS OF THE UNITED STATES

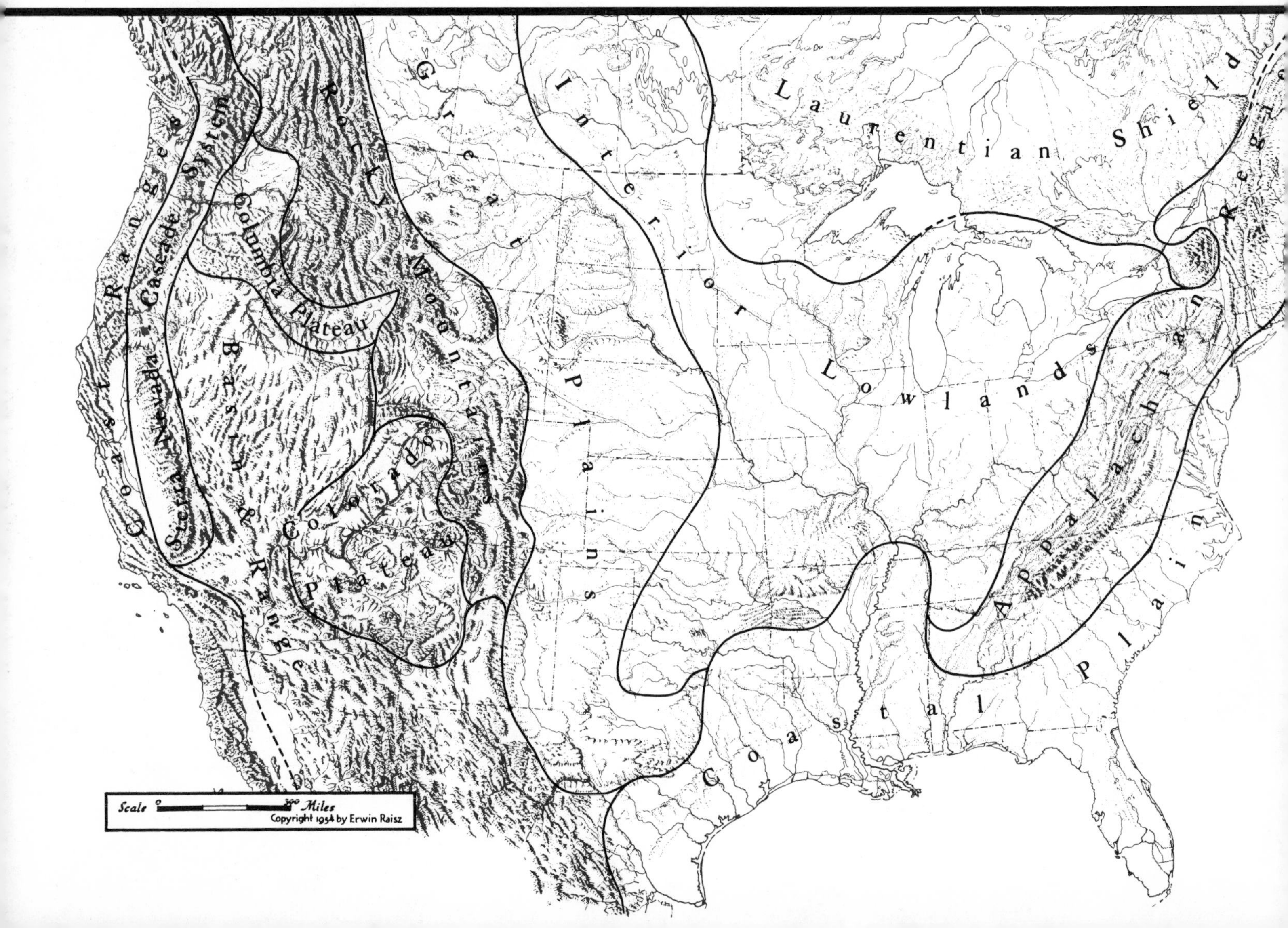

Coast Ranges
Sierra Nevada Cascade System
Columbia Plateau
Basin & Range
Colorado Plateau
Rocky Mountains
Great Plains
Interior Lowlands
Laurentian Shield
Appalachian Region
Coastal Plain

Scale 0 200 Miles

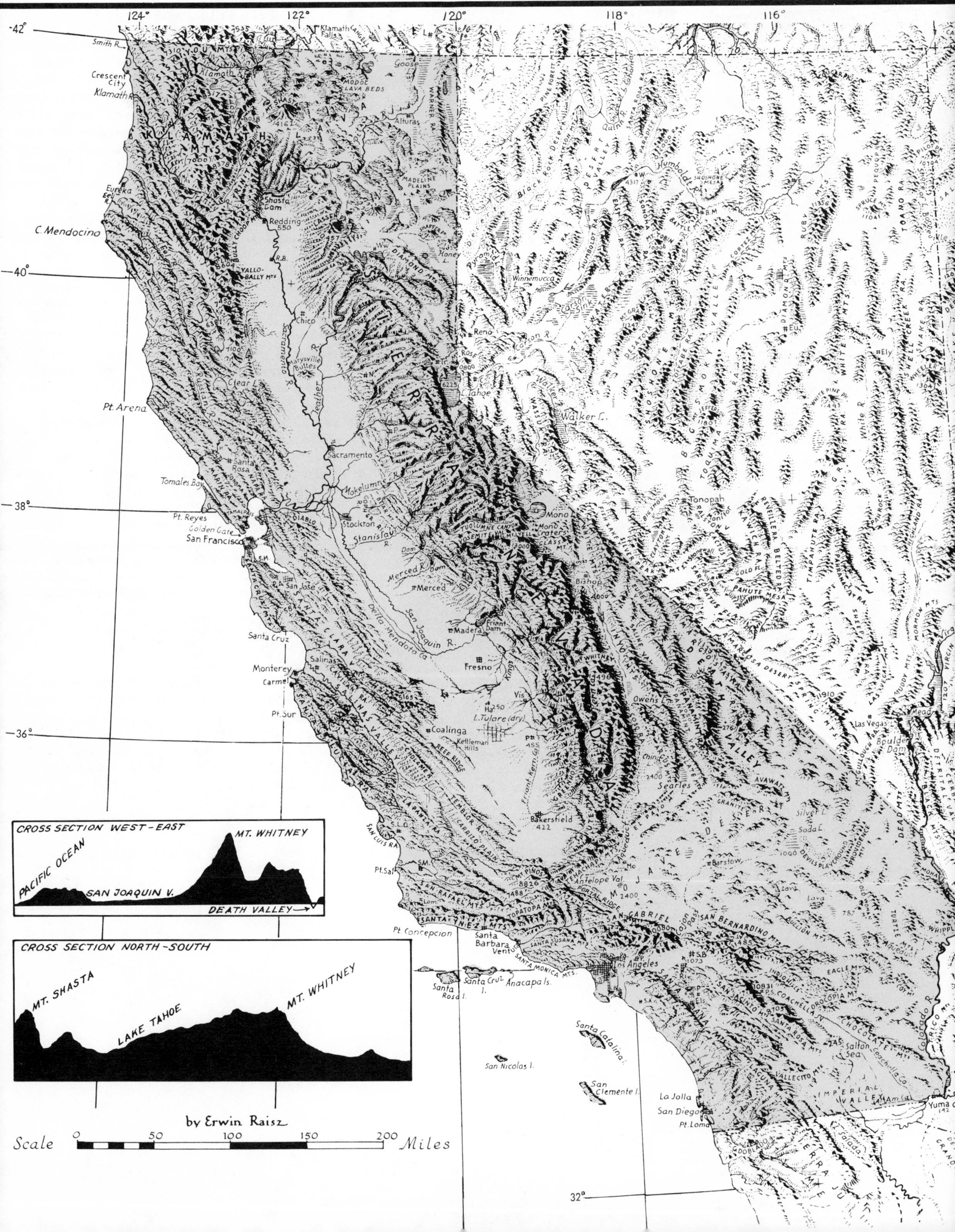
CROSS SECTION WEST-EAST
PACIFIC OCEAN
SAN JOAQUIN V.
MT. WHITNEY
DEATH VALLEY
CROSS SECTION NORTH-SOUTH
MT. SHASTA
LAKE TAHOE
MT. WHITNEY
by Erwin Raisz
Scale 0 50 100 150 200 Miles
Crescent City
Eureka
C. Mendocino
Pt. Arena
Tomales Bay
Pt. Reyes
Golden Gate
San Francisco
San Jose
Santa Cruz
Monterey
Carmel
Pt. Sur
Redding
Chico
Sacramento
Stockton
Merced
Madera
Fresno
Coalinga
Bakersfield
L. Tahoe
Reno
Winnemucca
Tonopah
Bishop
Mono L.
Walker L.
Las Vegas
Boulder Dam
Pt. Concepcion
Santa Barbara
Los Angeles
San Bernardino
Santa Catalina
San Nicolas I.
San Clemente I.
La Jolla
San Diego
Pt. Loma
Salton Sea
Imperial Valley
Yuma
Death Valley
Mojave Desert

LAND FORMS

Differences in the landscape are produced by differences in rock type, in the kinds of structures created by earth movement and in the erosional processes which tend to modify and sculpture the land. Different parts of California have had varying geologic histories with the result that there is a great diversity in the appearance of the landforms of the state.

The *Sierra Nevada* is a single mountain range nearly as extensive as the Alpine system of Europe and occupying one-fifth the total area of California. It varies from 40 to 80 miles in width and extends some 430 miles in a north south direction. The range represents a huge granite block which has been pushed up from the interior of the earth and tilted to the west.

The *Cascade Range* and the *Modoc Plateau* are of volcanic origin. The Cascades consist of innumerable volcanic peaks and lava flows of variable age and degree of erosion. The Modoc Plateau represents an extension of the Columbia Lava Plateau into northeastern California. It has an undulating surface with an average elevation of about 4,500 feet on top of which a number of hilly and mountainous areas are superimposed.

The *Klamath Mountain Province* is an old land mass consisting of modified sedimentary and igneous rock which has been uplifted, folded, faulted and eroded a number of times. Its rugged mountains attain elevations of 8,000 to 9,000 feet.

The *Coast Ranges* are lower and younger than the Klamath Mountains. They extend in a northwest-southeast direction and are composed chiefly of sedimentary rocks.

The only major east-west trending mountain ranges in California are included in the *Transverse Range* which gains its name from this fact. These ranges represent a complex set of fold-fault mountains and valleys containing a wide variety of rock types.

The *Peninsular Ranges* lie to the south of the Transverse Ranges and may be compared to the Sierra Nevada in terms of their origin and the rock type of which they are composed. For the most part they are fault-block mountains with steep eastern faces and with granitic cores. Marine terraces consisting of poorly consolidated sedimentary material cover their western slopes.

Although the *Colorado Desert Landform Province* lies in the trough which continues southward as the Gulf of California, it is doubtful whether it was covered by the ocean in recent time. Well corings reveal only continental deposits and indicate that the area was occupied by fresh water lakes in recent geologic time.

The *Mojave Desert* and the *Basin and Range* provinces have somewhat similar geologic histories. However, the mountains and valleys in the Basin and Range Province trend in a N — S direction while much of the western portion of the Mojave Desert is a plain upon which a few isolated mountains appear.

The *Central Valley* of California is a large alluvial plain filled with the sediments carried out of the mountains enclosing it by the myriad numbers of streams tributary to the San Joaquin and Sacramento rivers.

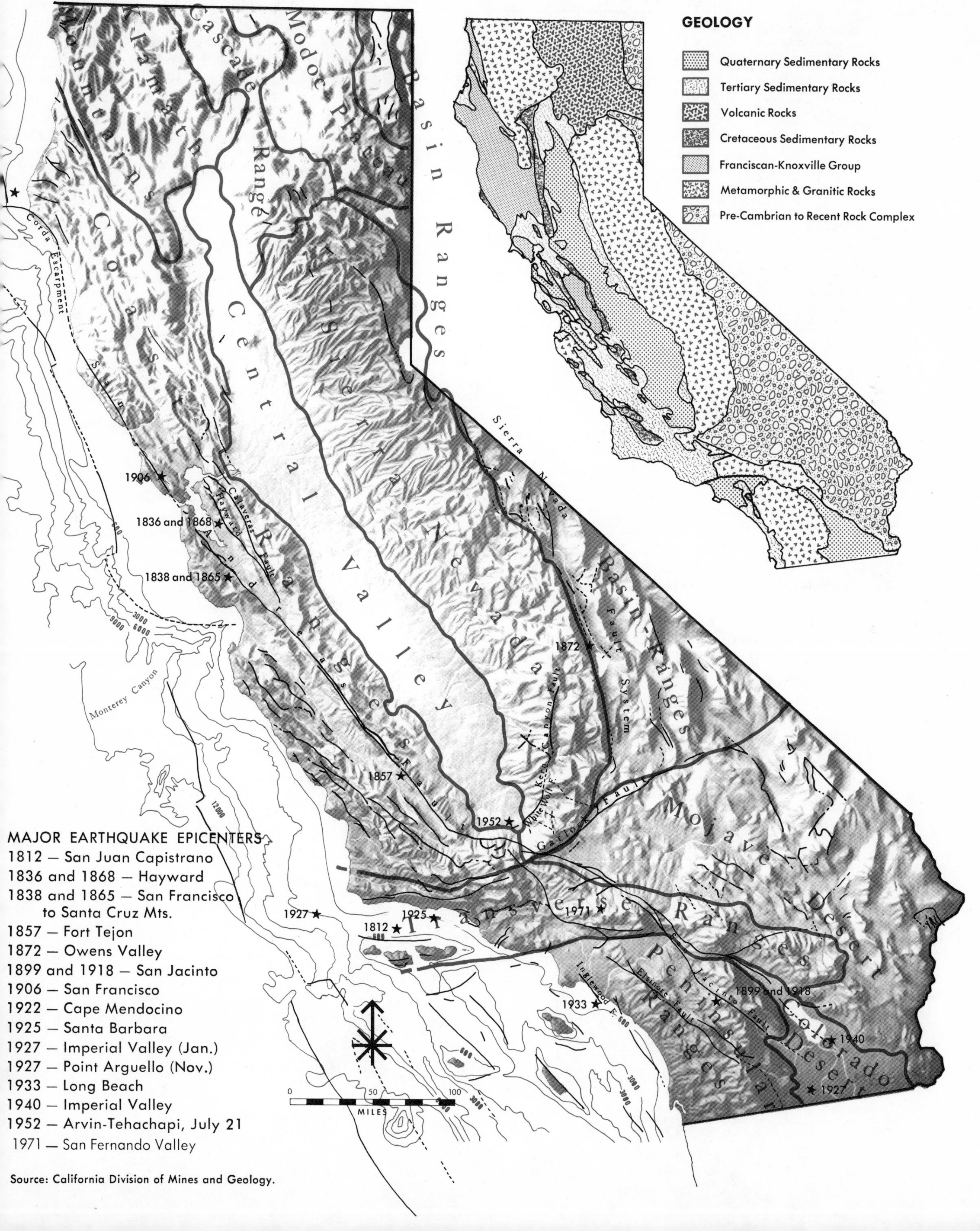

Source: California Division of Mines and Geology.

RIVERS OF CALIFORNIA

Hydrographic area River (or other stream) and Station	Drainage area (in sq. mi.)	50-year mean ° (Acre-feet)
North Coastal Area		
Trinity at Lewiston	727	1,195,400
Klamath, Copco to Somesbar	4,400	4,433,900
Eel at Scotia	3,113	5,157,500
Russian at Healdsburg	791	800,800
San Francisco Bay Area		
Napa near St. Helena	81	61,650
Coyote Creek near Madrone	194	49,620
Central Coastal Area		
Arroyo Seco near Soledad	241	117,450
Nacimiento below Nacimiento Dam, near Bradley	322	185,000
South Coastal Area		
Sespe Creek near Fillmore	254	65,320
Arroyo Seco near Pasadena	16	6,220
Santa Ana near Mentone	202	64,880
Sacramento Valley Area		
Sacramento, Inflow to Shasta	6,665	5,483,400
Sacramento near Red Bluff	9,300	7,953,800
Feather near Oroville	3,632	4,350,100
Yuba at Smartville	1,104	2,273,100
Bear near Wheatland	295	323,100
American, Inflow to Folsom	1,889	2,636,400
Cosumnes at Michigan Bar	537	358,900
Mokelumne, Inflow to Pardee	538	722,100
San Joaquin Valley Area		
Stanislaus, Inflow to Melones	905	1,111,000
Tuolumne, Inflow to Don Pedro	1,534	1,802,600
Merced, Inflow to Exchequer	1,029	942,600
San Joaquin, Inflow to Friant	1,675	1,703,700
Kings, Inflow to Pine Flat	1,542	1,607,300
Kaweah near Three Rivers	561	393,650
Tule, Inflow to Success	393	133,000
Kern near Bakersfield	2,420	687,530
Lohontan Area		
Truckee, Tahoe to Farad	421	388,600
Owens, below Long Valley	437	155,100

HYDROGRAPHY

Surface water resources have always been of great significance to Californians. Indian villages and Spanish presidios, pueblos and missions were dependent on supplies drawn from easily obtained surface sources. Gold was discovered in the South Fork of the American River; water was essential in the mining of gold.

A map of the rivers and lakes in California constantly changes as man modifies the landscape. In the past, several large lakes occupied the southern end of the Central Valley; steamboats operated across both Tulare Lake and Owens Lake. These lakes now appear only in unusually wet periods for man has built dams across the rivers which formerly supplied the lakes with water and thus has created new lakes in new places.

The estimated mean seasonal runoff of all California streams is about 71,000,000 acre-feet. (An acre-foot of water is enough to cover an acre of land with water one foot deep and is equal to 325,851 gallons.) The streams of the North Coastal Area furnish about 41 per cent of the total for the State, and the streams of the Sacramento River Basin furnish about 32 per cent. Most of the remainder of the water is in the San Joaquin Valley.

The streamflow maps of California and the United States give one an idea of the relative amounts of water available in each significant stream. They also indicate the relative quantities of water to be found in the various parts of the state and of the United States.

AVERAGE ANNUAL STREAMFLOW

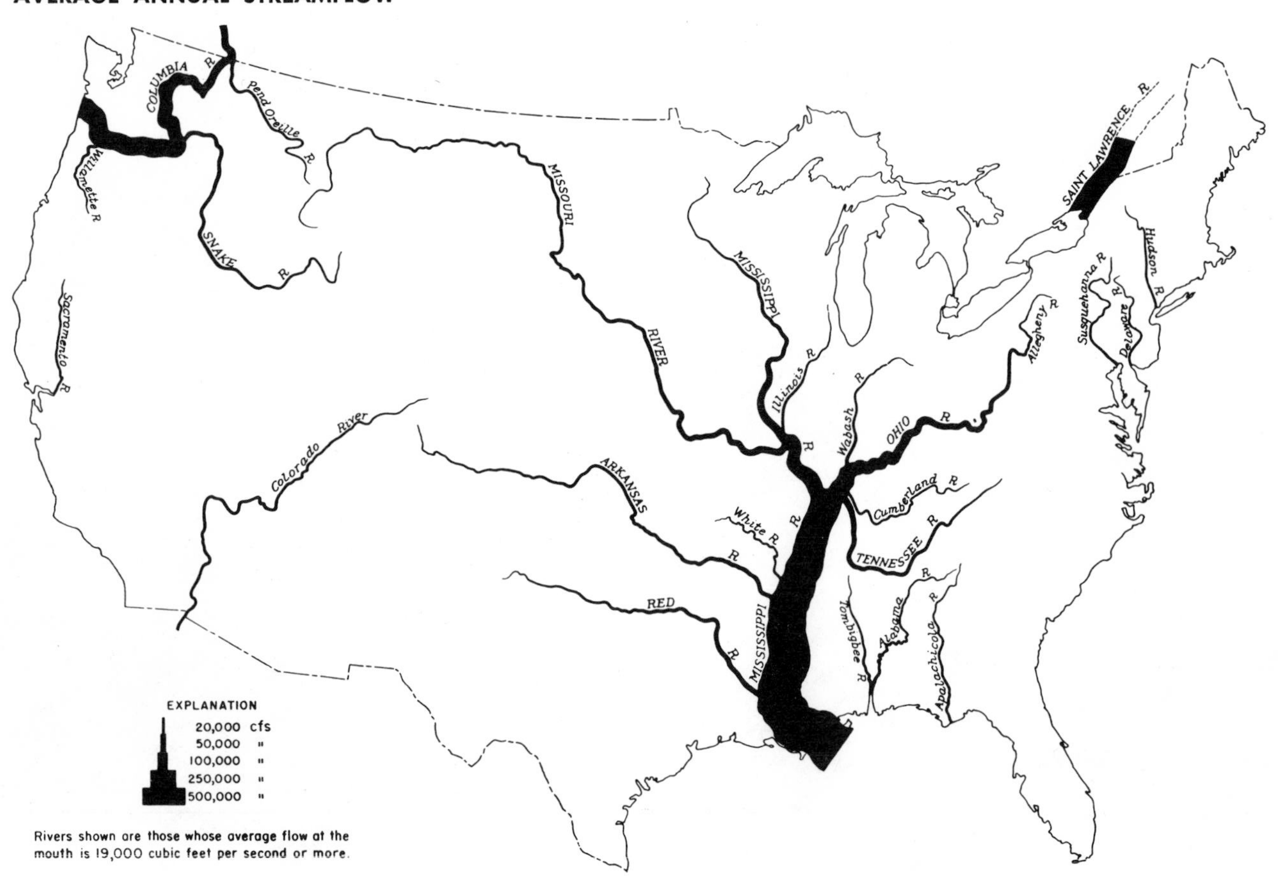

Rivers shown are those whose average flow at the mouth is 19,000 cubic feet per second or more.

AVERAGE ANNUAL STREAMFLOW
100,000-500,000 ACRE-FEET
OVER 500,000 ACRE-FEET
OREGON
NEVADA
ARIZONA
MEXICO
PACIFIC OCEAN
Shasta Lake
Eagle Lake
Honey Lake
Goose Lake
Lake Almanor
Clear Lake
Lake Tahoe
Mono Lake
Owens Lake
Tulare Lake
Lake Mead
Boulder Dam
PARKER RESERVOIR
LOS ANGELES AQUEDUCT
COLORADO RIVER AQUEDUCT
SAN DIEGO AQUEDUCT
COACHELLA CANAL
ALL AMERICAN CANAL
SAN FRANCISCO
OAKLAND
SACRAMENTO
STOCKTON
LOS ANGELES
SAN DIEGO
0 50 100
MILES

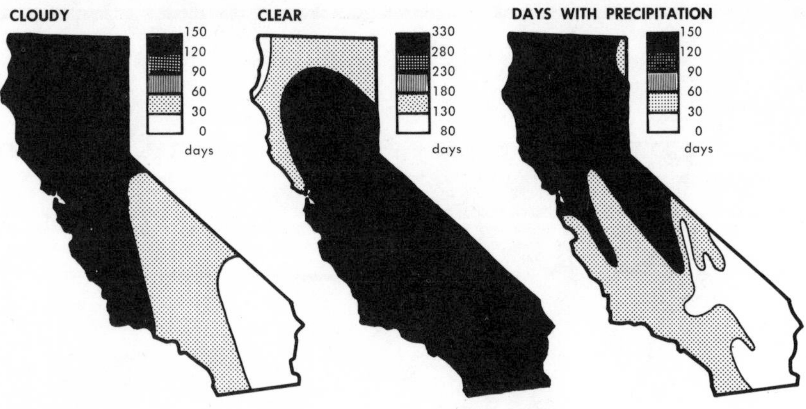

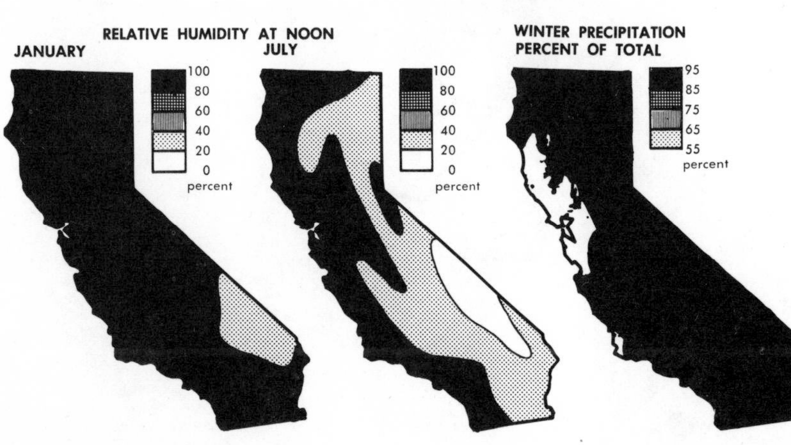

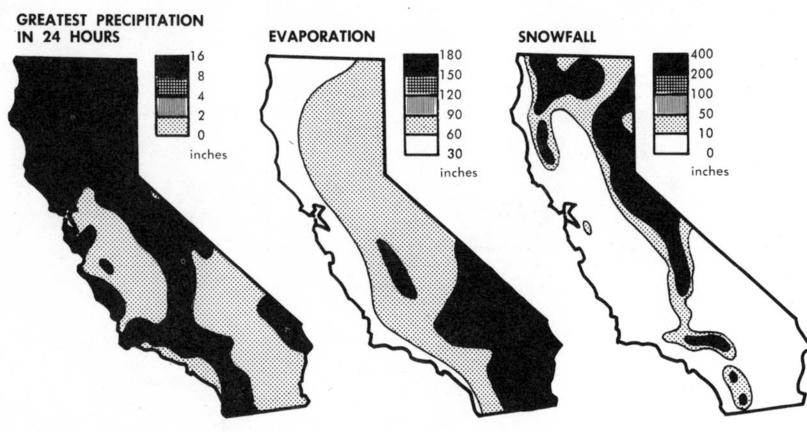

Source: U.S. Weather Bureau.

PRECIPITATION

A major portion of the precipitation which occurs in California is associated with cyclonic storms which sweep out of the Gulf of Alaska during the winter season to deposit their load of snow and rain on the mountains and valleys of this part of the United States.

In the fall months these cyclonic storms swing farther and farther to the south and bring wind and rain to northern California. In winter their centers may pass over the California coastline and on rare occasions may even pass over southern California. In general, the amount of precipitation which falls at any given locality is proportional to the distance to the center of the low. The nearer one is to the center, the greater is the precipitation.

Three basic generalizations may be derived from an examination of the map of mean annual precipitation. They are: (1) precipitation generally decreases from north to south and west to east except where mountains intervene; (2) precipitation increases with elevation up to a certain point; (3) precipitation is greater on the west and south sides of the mountains than on the north and east sides.

Northern California receives more rain than southern California because it is affected by more cyclonic storms, and the centers of these storms generally are nearer to northern California than to southern California. The first cyclonic storms bring rain to northern California in late September or October and to southern California by late October or November. They continue to bring clouds, wind and rain to the state until late April or May and may even affect northern California in June.

The increase of precipitation with elevation is a reflection of the fact that air at higher levels along a mountain slope has been cooled a good deal more than at lower levels, and that a greater thickness of air is involved in the lifting and cooling process. As the lifting and cooling process is continued, a point of diminishing returns is reached because of the fact that the bulk of the water has been precipitated from the air.

The elevation at which maximum precipitation occurs during any particular storm is dependent on a number of factors. The most significant of these is the total amount of water in the air.

Because of the counter-clockwise flow of air around cyclonic storms in the northern hemisphere, the moisture-bearing winds in advance

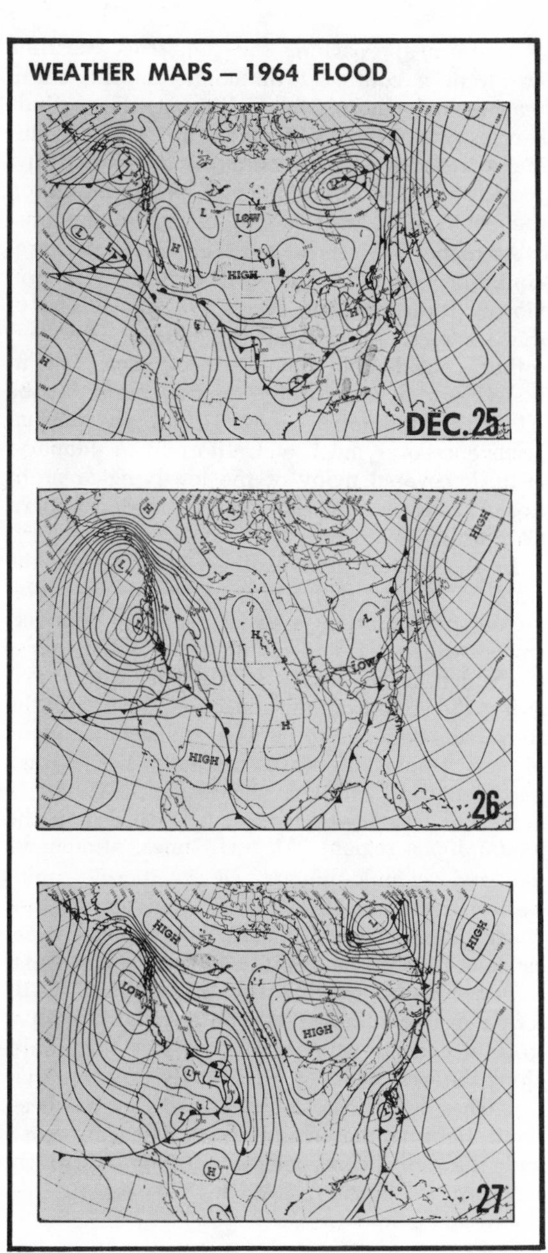

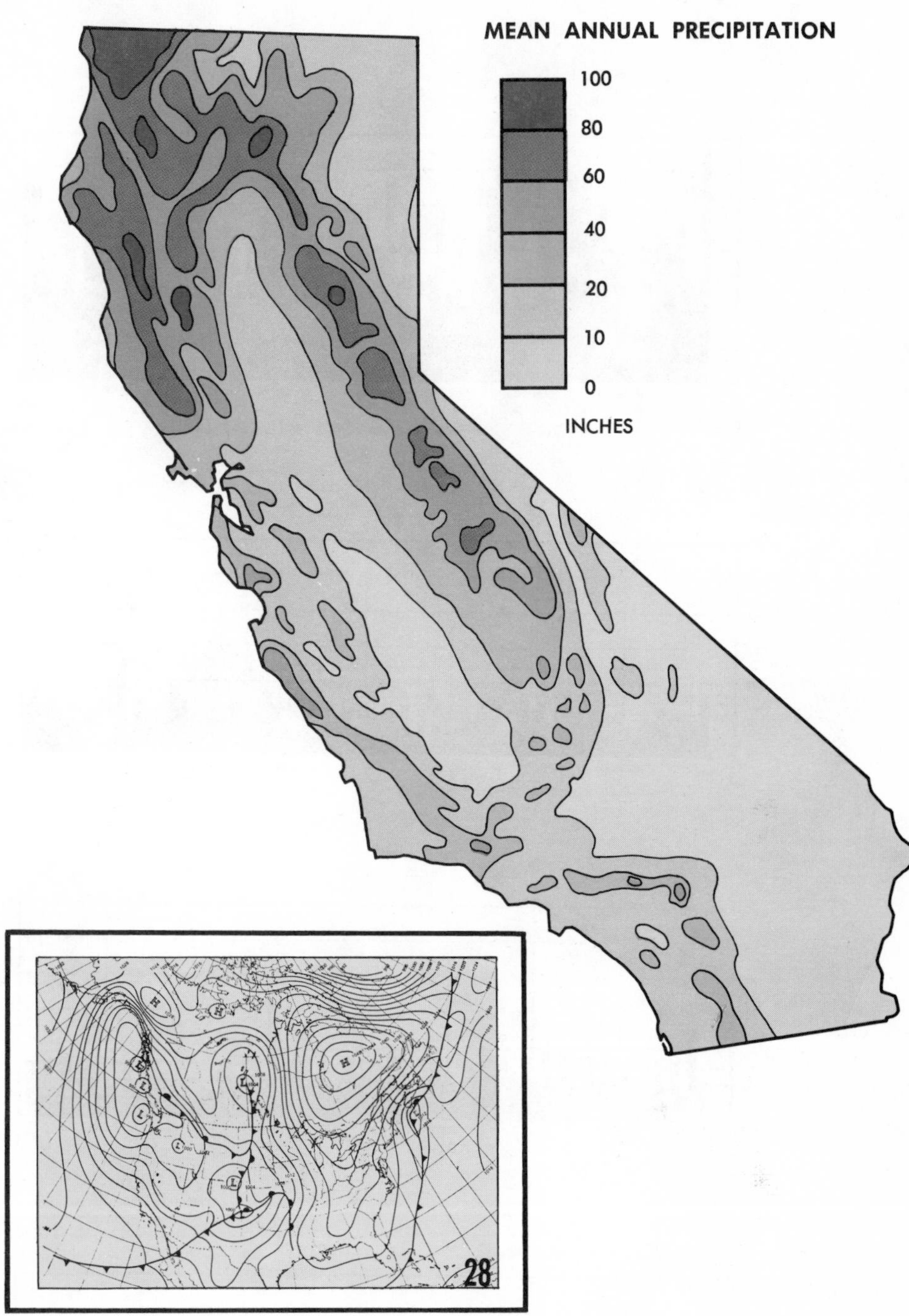

WEST-EAST CROSS SECTION

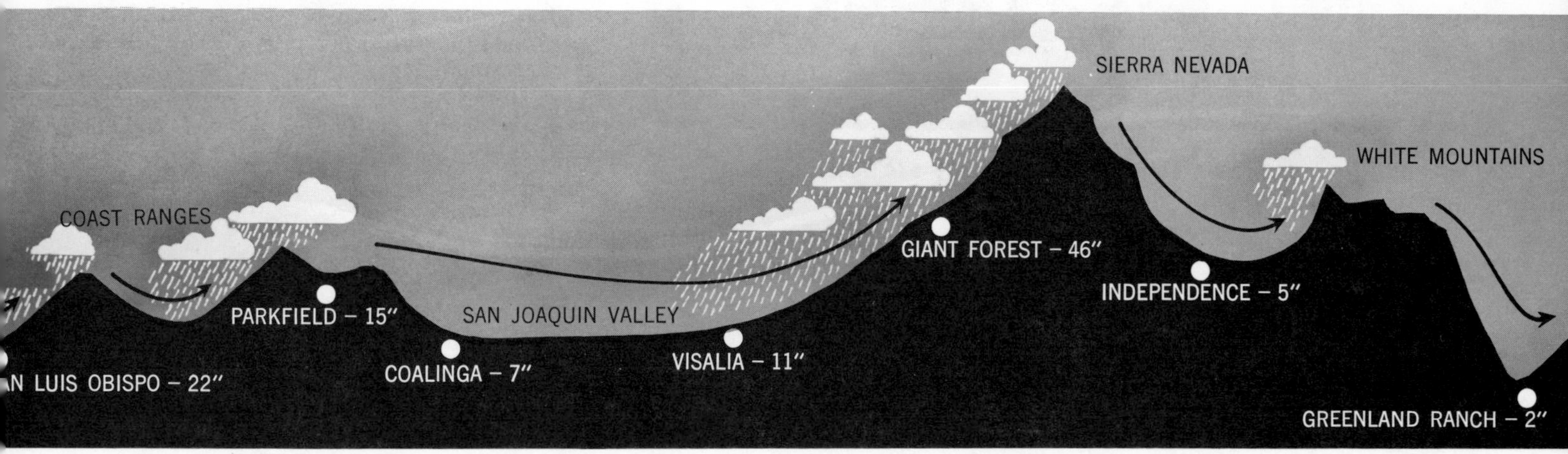

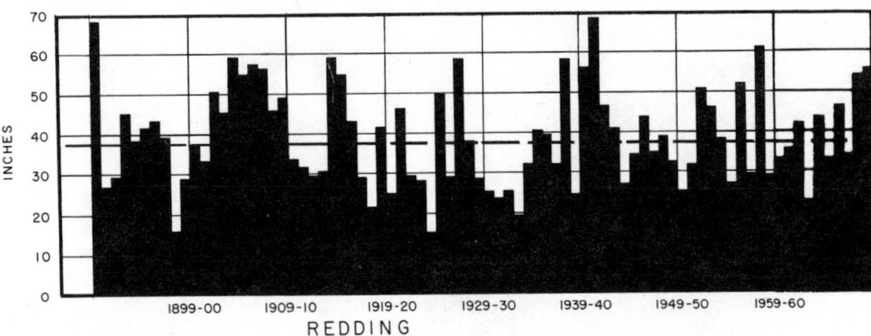

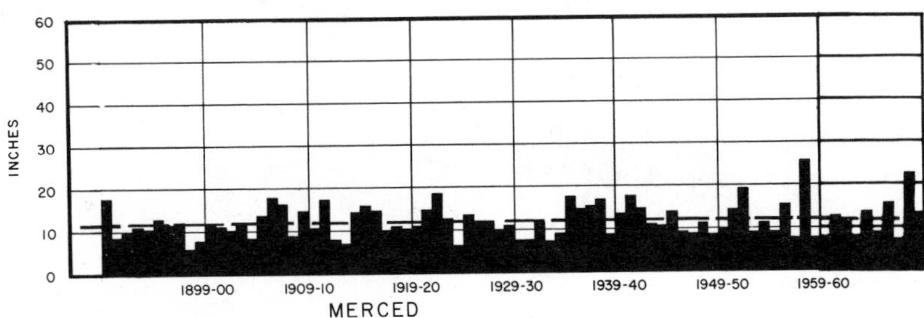

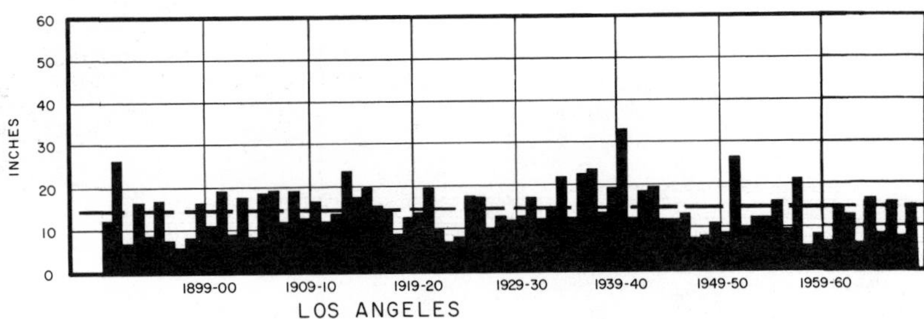

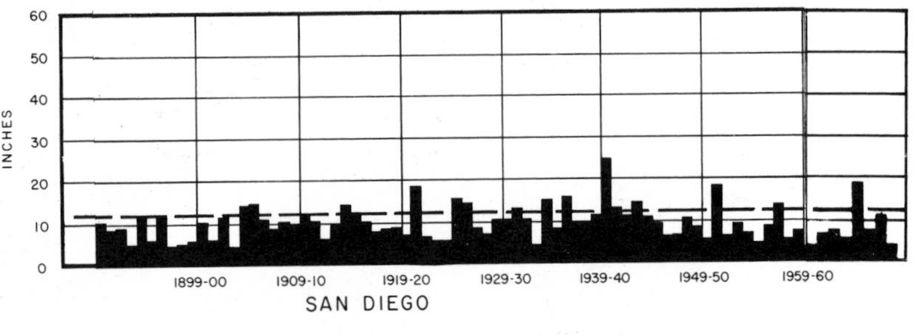

PRECIPITATION AT REPRESENTATIVE STATIONS

of a storm approaching the California coastline are from a southerly direction. As the storm reaches and passes over California, these winds become westerly. As air piles up on the south and west sides of the mountains it is lifted and cooled, with resultant heavier precipitation on the south and west slopes of the mountains.

At higher elevations, most of the winter precipitation occurs in the form of snow. As previously indicated, the Sierra Nevada is one of the snowiest regions in the world. However, rather considerable quantities of snow fall in the Cascades and in the higher parts of the Klamath Mountains. Snow is not an unusual occurrence over much of California; in January, 1949, it covered many of the low-lying areas of Los Angeles as well as other parts of southern California.

In the eastern and southeastern part of the state, a considerable proportion of the total annual precipitation is associated with thunderstorms which form in a tongue of warm, moist air which originates over tropical waters and enters the state from the southeast. These are the so-called "Sonora Storms". At times the tongue of moist tropical air which produces these storms may penetrate as far as the Klamath Mountain area, and thunderstorms will be recorded in the mountainous regions. At such times, alto-cumulus, and cumulo-nimbus clouds permit midwesterners to reminisce about summer weather "back home". Such storms are the principal cause of fires in the forests of the Sierra Nevada.

Occasionally, precipitation in southern California in late summer may be associated with a tropical cyclone which works its way northward along the west coast of Mexico. By the time that it reaches San Diego County, its energy has been spent fighting the mountainous Mexican coast, and the only consequences of the storm in southern California are high temperatures, cloudy skies and rain.

TEMPERATURE

The maps of average temperature on these pages portray in broad detail the climatic conditions which prevail in each part of California. The coastal regions and the valleys leading into the interior are bathed in cool sea air during the summer months while the interior portions of California swelter under high temperatures. In winter the ocean remains warmer than the land and coastal locations have higher maximum temperatures than the interior.

TEMPERATURE

Low clouds and fog along the coast during summer and in the Central Valley during winter play a role in limiting temperature variation between day and night.

However, great extremes of temperature have been recorded in the state. On July 10, 1913 a temperature of 134° F was recorded at Greenland Ranch. Each summer temperatures above 100° F are observed in the desert regions while each winter temperatures below zero occur in the High Sierra and in the Modoc Plateau Area. Even the usually mild coastal zones record extreme temperature conditions from time to time. Heat waves usually occur in late summer and fall while freezes involving the influx of cold air from Canada occur in December and January.

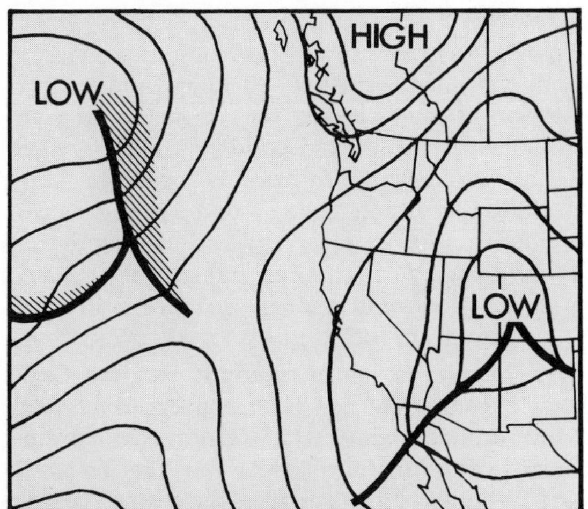

REPRESENTATIVE FREEZE MAP

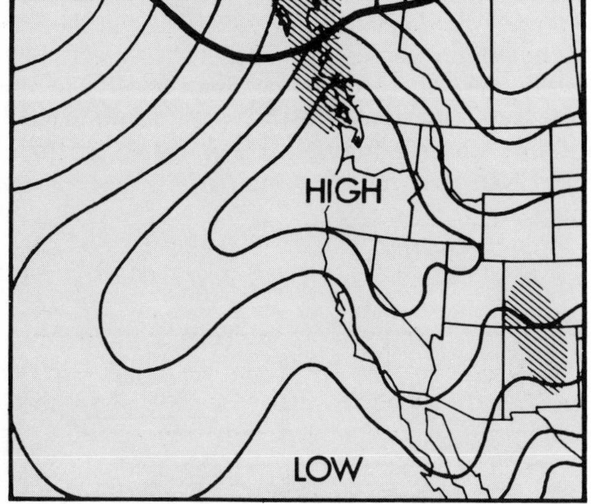

HIGH TEMPERATURE CONDITIONS

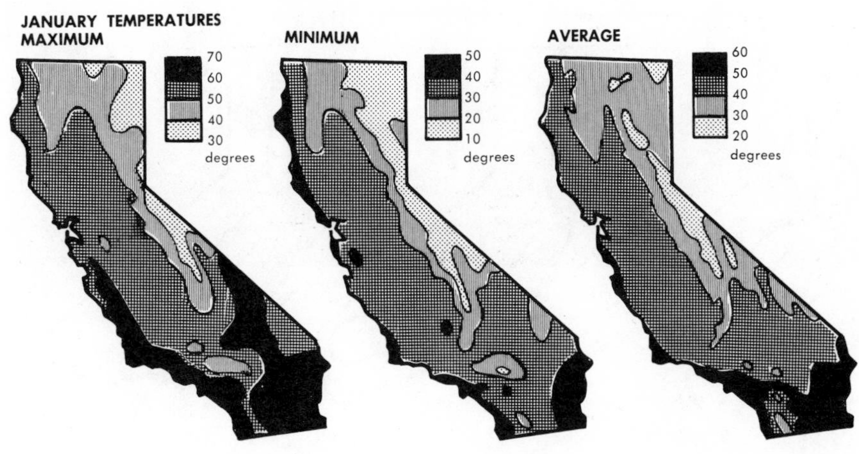

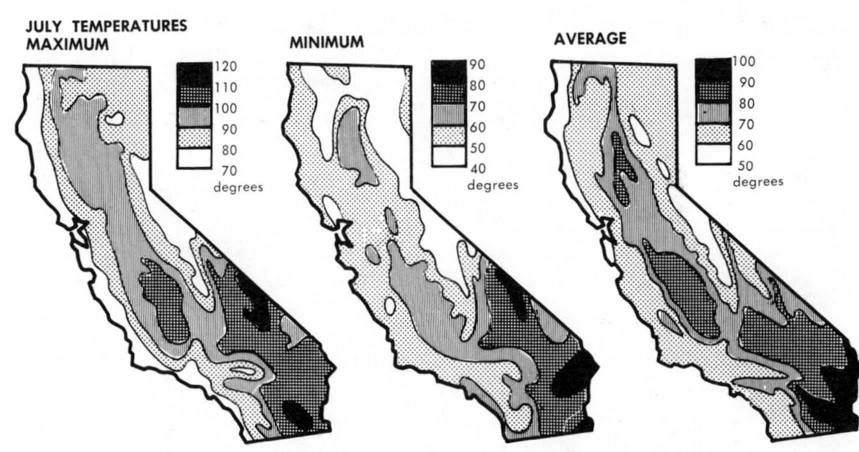

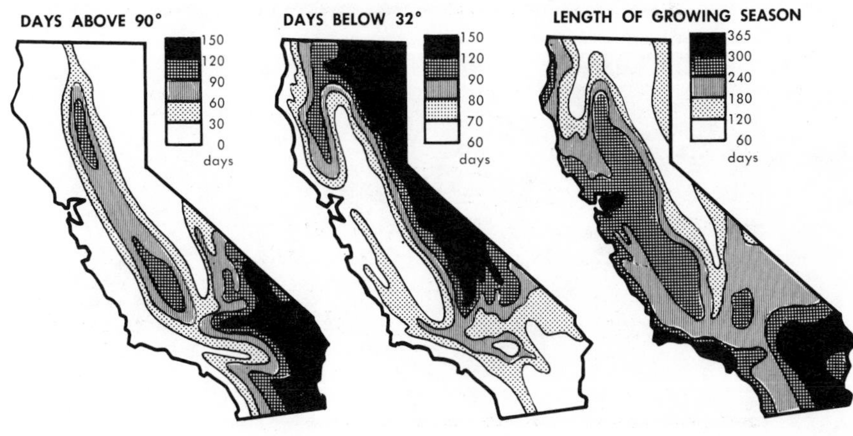

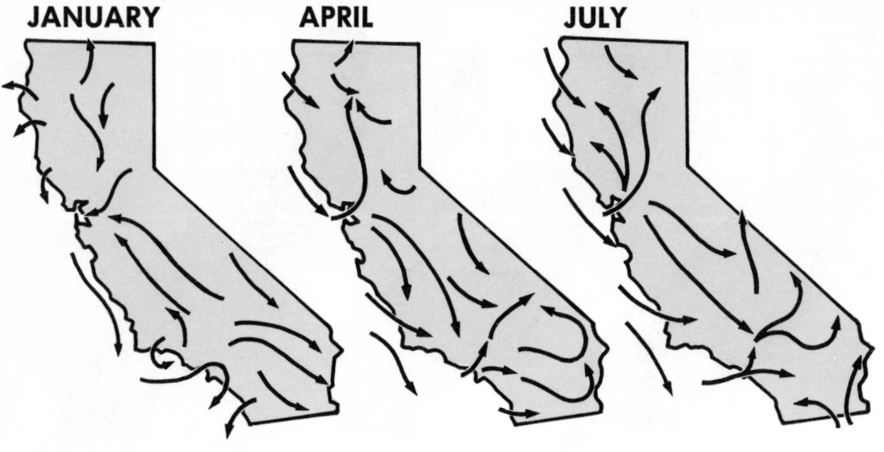

SUMMER WEATHER CONDITIONS

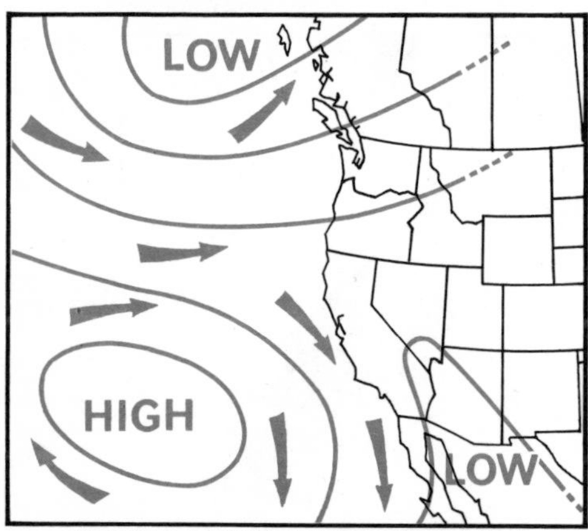

TYPICAL SUMMER WEATHER MAP

TOPOGRAPHY AND WIND FLOW PATTERNS

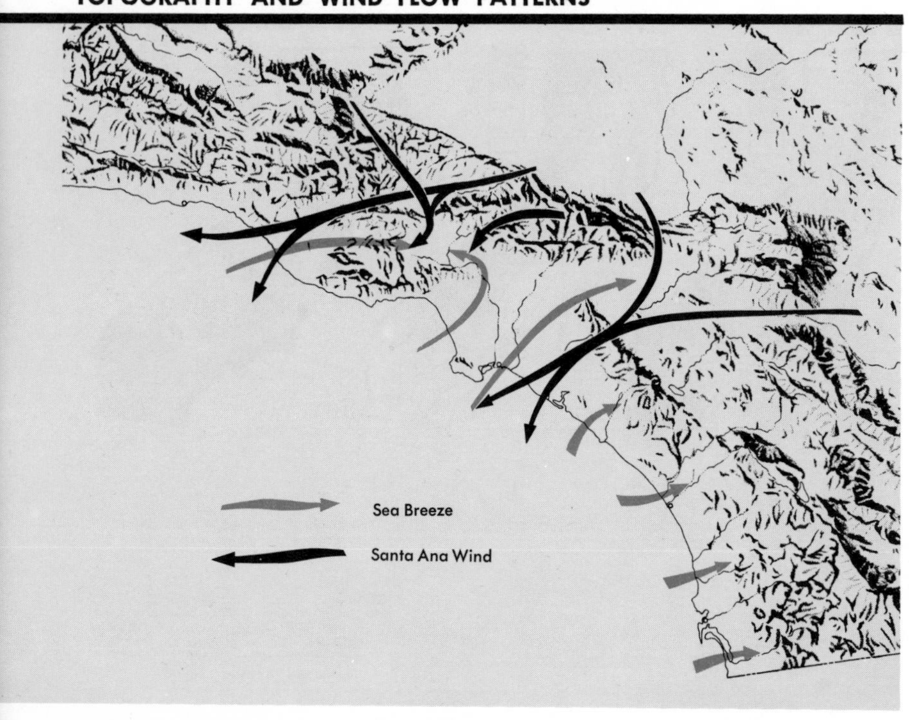

WIND

During the warm season, the pressure gradient over the southwestern part of the United States is generally weak, and wind movement is generally light. The dominant movements of air in coastal areas are those associated with the land and sea breeze while in mountainous and hilly regions, mountain and valley breezes prevail. These movements of the lower layers of the atmosphere are augmented and modified by the general flow of air around the Pacific High and into the Thermal Low.

The sea breeze generally sets in between 9 and 10 A.M. and blows strongly during the daylight hours as the colder and denser sea air moves inland to displace the heated and less dense air of the interior.

Winds associated with cyclonic storms may attain velocities of 40 to 60 knots and do considerable damage along the coast. During the days of sailing ships and paddle wheel steamers, a large number of shipwrecks occurred along the California coast. Today's damage is generally confined to the toppling of shallow-rooted trees which in turn damage other objects. The highest wind velocities and greatest turbulence are generally found in the vicinity of the cold front.

The intensification of pressure over the Great Basin, accompanied by the formation of a trough of low pressure along the Californa coast line, results in the flow of air from the interior to the coast. Whether this air arrives as a warm or cold wind, depends upon its initial temperature and the rate of flow from the interior. The pressure gradient between the interior and the coast determines whether the flow of air is confined to the mountain passes or whether it passed completely over the mountains themselves. Such a flow has been referred to as a "foehn wind," and in California they have been called northers, easterlies, Santyanas, Santa Anas and desert winds.

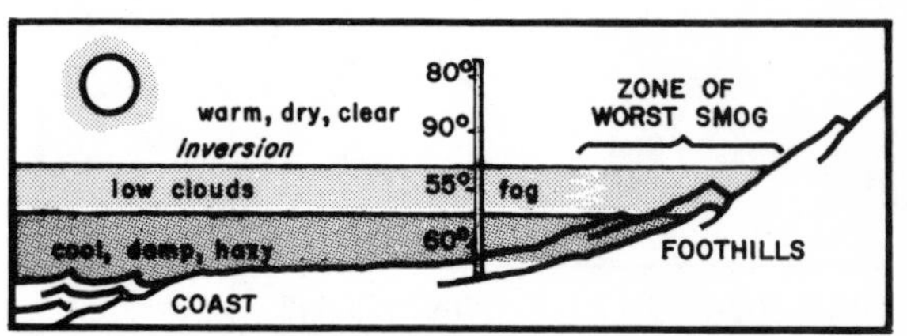

SMOG CONDITIONS

Stations Listed in the Appendix
Class A Weather Stations
CRESCENT CITY
MOUNT SHASTA
ALTURAS
EUREKA
SCOTIA
REDDING
SUSANVILLE
RED BLUFF
CHICO
FORT BRAGG
OROVILLE
BLUE CANYON
TAHOE
MARKLEEVILLE
SACRAMENTO
DAVIS
SANTA ROSA
BRIDGEPORT
STOCKTON
OAKLAND
SAN FRANCISCO
SAN FRANCISCO AIRORT
MODESTO
YOSEMITE NATIONAL PARK
SAN JOSE
BISHOP
MERCED
FRESNO
SALINAS
GIANT FOREST
GREENLAND RANCH
KING CITY
COALINGA
PORTERVILLE
PT. PIEDRAS BLANCAS
PASO ROBLES
BAKERSFIELD
SAN LUIS OBISPO
SANTA MARIA
BARSTOW
NEEDLES
SANDBERG
PALMDALE
VICTORVILLE
POINT ARGUELLO
SANTA BARBARA
SAN FERNANDO
BURBANK
BIG BEAR
OXNARD
SAN BERNARDINO
SANTA MONICA
LOS ANGELES
RIVERSIDE
LOS ANGELES INT. AIRPORT
LONG BEACH
SANTA ANA
PALM SPRINGS
INDIO
BLYTHE
SANTA CATALINA
SAN NICHOLAS IS.
ESCONDIDO
SAN DIEGO
PACIFIC OCEAN
0 50 100
MILES

16 CLIMATIC REGIONS

CLIMATE

The climate of California and of various parts of California has long been the subject of debate among the citizens of the state and the nation. Does Florida really have a better winter climate than southern California? Are the chilling fog and winds of San Francisco to be preferred to the heat and smog of Los Angeles?

Regardless of one's opinions on the above questions, the fact remains that in California one does have the choice of a number of climatic regions in which to reside. There are many more choices available within the confines of the state than in virtually any other comparable portion of the globe.

A system of classifying climates used in all parts of the world is that devised by the German climatologist, Wladimir Köppen.

The map at the bottom of this page represents a simplification of the map developed by Köppen. On it the major climatic regions of the world are shown. The *A* climates are regions which are constantly hot and humid while the *B* climates represent the steppe and desert regions of the world. Within the *C* and *D* categories are regions with cool or cold winters, warm to hot summers and sufficient precipitation to support plant growth. The polar climates are shown by the symbol *E*. Areas with climates similar to those found in much of California are called Mediterranean climates because of their occurrence around the sea of the same name. They may also be found in central Chile, South Africa and in Australia. They are identified by the symbol Cs.

The map of California climates shows the principal sub-types of Köppen's classification found in this state. In adapting the system to this continent American geographers have suggested four new sub-types and additional modifications of some of the original categories. Some of these are shown on the map on the opposite page.

The insets on the map of California climates show average rainfall and temperature conditions for typical stations in each climatic region for the California rainfall season which extends from July 1 to June 30. They show the contrasting climatic conditions found in the various parts of the state.

CLIMATES OF THE WORLD

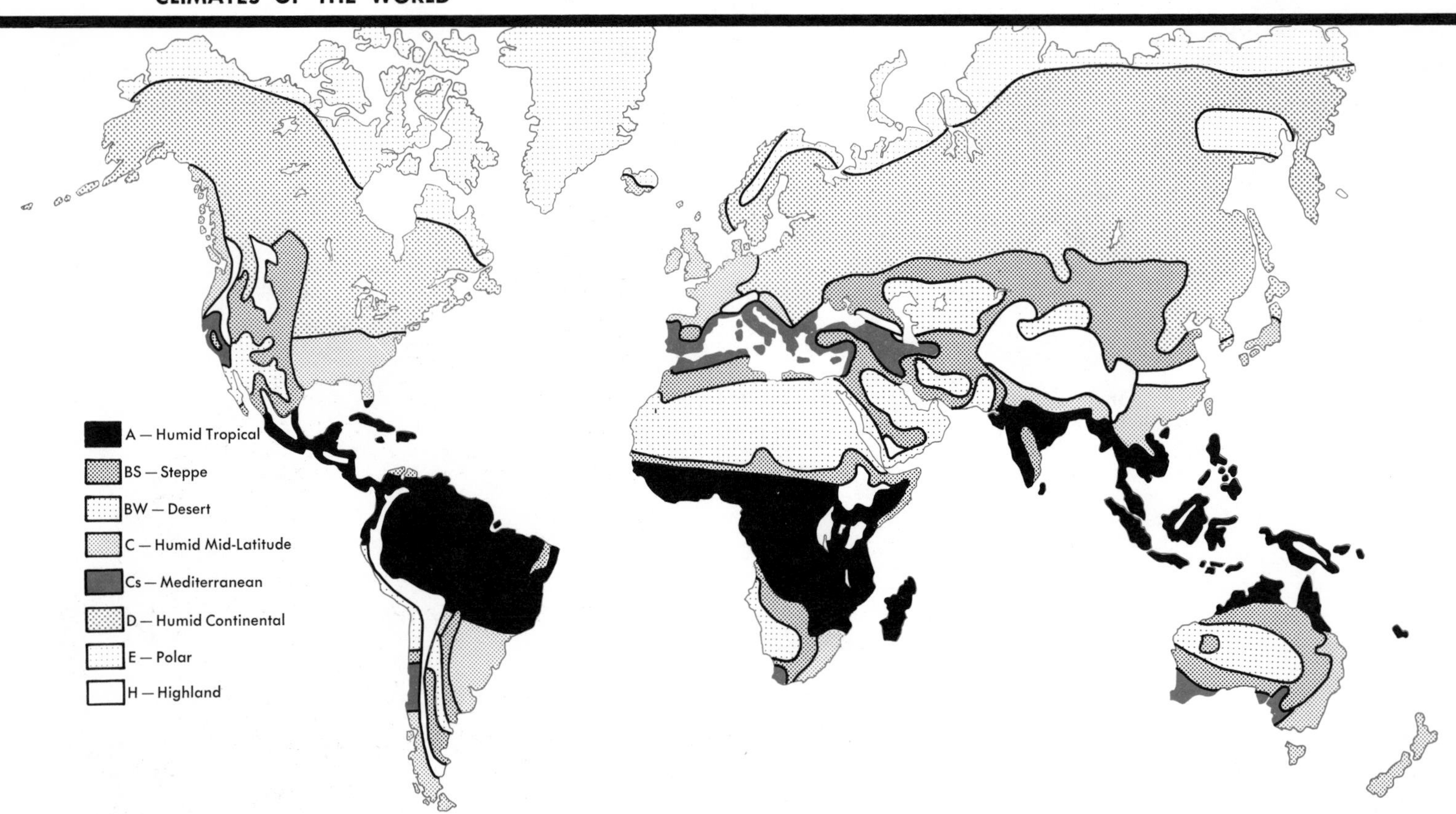

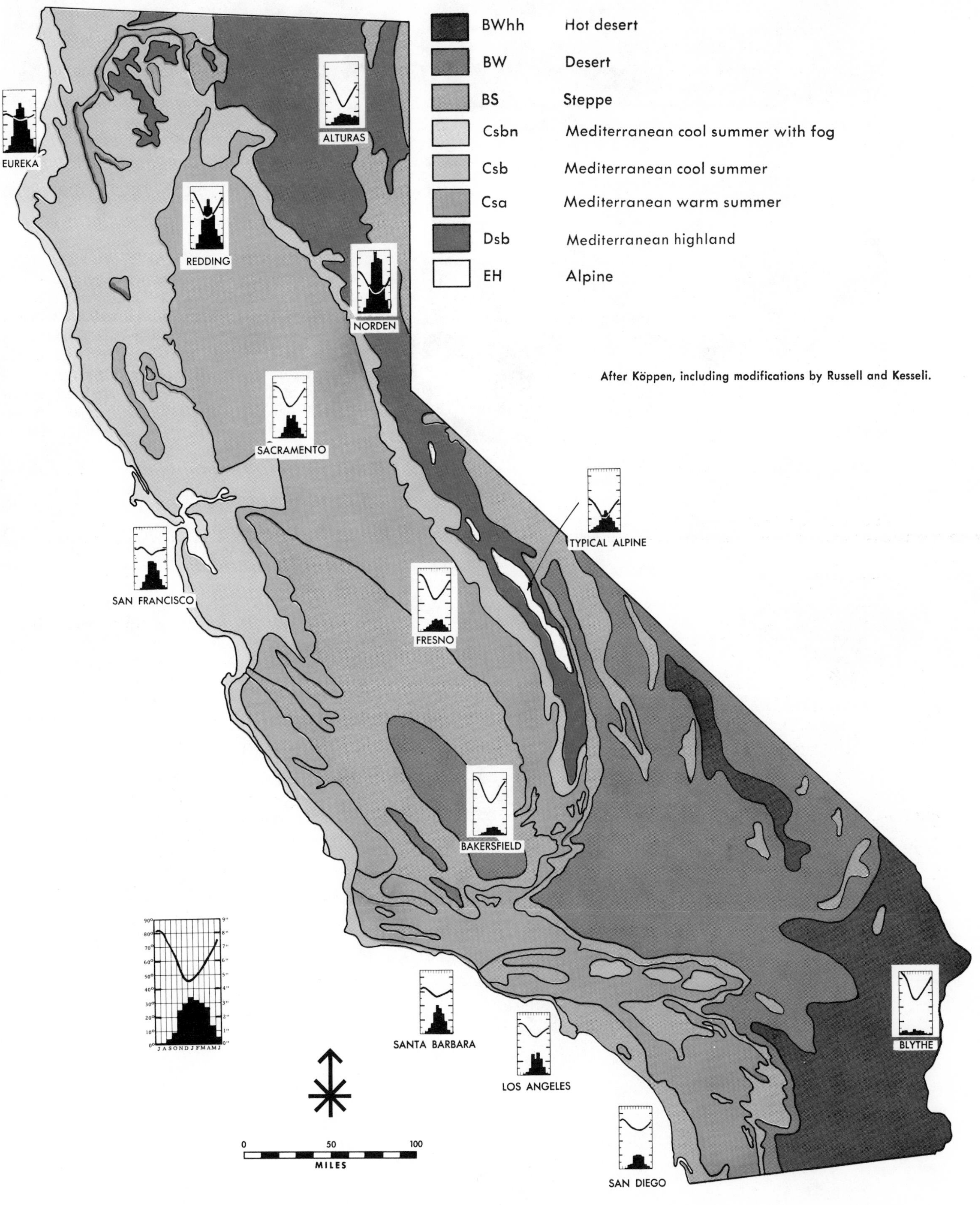
BWhh Hot desert
BW Desert
BS Steppe
Csbn Mediterranean cool summer with fog
Csb Mediterranean cool summer
Csa Mediterranean warm summer
Dsb Mediterranean highland
EH Alpine
After Köppen, including modifications by Russell and Kesseli.
EUREKA
ALTURAS
REDDING
NORDEN
SACRAMENTO
TYPICAL ALPINE
SAN FRANCISCO
FRESNO
BAKERSFIELD
SANTA BARBARA
LOS ANGELES
SAN DIEGO
BLYTHE
J A S O N D J F M A M J
0
50
100
MILES

Woodland grass landscape in the Coast Ranges.

Desert shrubs.

LIFE ZONES

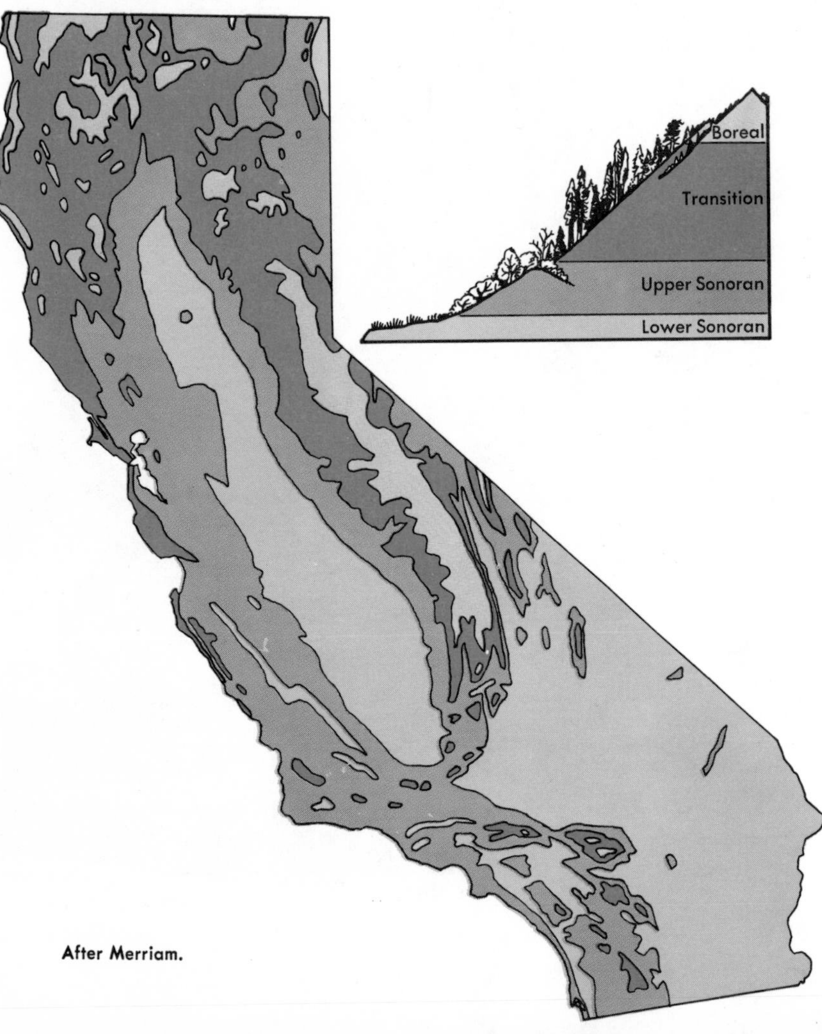

After Merriam.

The redwood forest in the Eel River Valley.

NATURAL VEGETATION

Prior to the coming of the white man, portions of the California landscape had an entirely different appearance than they do today. In many valley areas, stately oaks lifted their heads above the Indians and the wild animals crossing the valley floor. Gallery forests extended along the streams of the Central Valley, and great tule marshes occupied the delta area of the Sacramento and San Joaquin rivers and the Tulare Basin. The eucalyptus was nowhere to be seen, and redwood forests covered mountain slopes in the counties surrounding San Francisco.

In spite of the work of man in modifying the landscape, the general arrangement of the larger associations of wild plants remains fairly much the same as it was in the days when the Spanish first came to California. The geographic distribution of living forms is governed by a number of different factors such as humidity, rainfall, temperature, altitude, solar radiation, soil and availability of plant nutrients. Plants possess tolerance for each of the factors necessary for plant growth. For each plant there are minimum and maximum values for each requirement which will limit its growth.

Within any given area in California there are generally a large number of different species of plants growing. Hence, any attempt to map the distribution of plants must involve generalizations based upon the principal species found in any given area. On the opposite page a number of different plant associations are depicted as they appeared in the pre-European period while the map of life zones on this page represents an attempt to identify major areas having similar environmental conditions, and, hence, similar groups of plants and animals.

Source: Wieslander and Jensen, *Forest Areas, Timber Volume and Vegetation Types in California.*

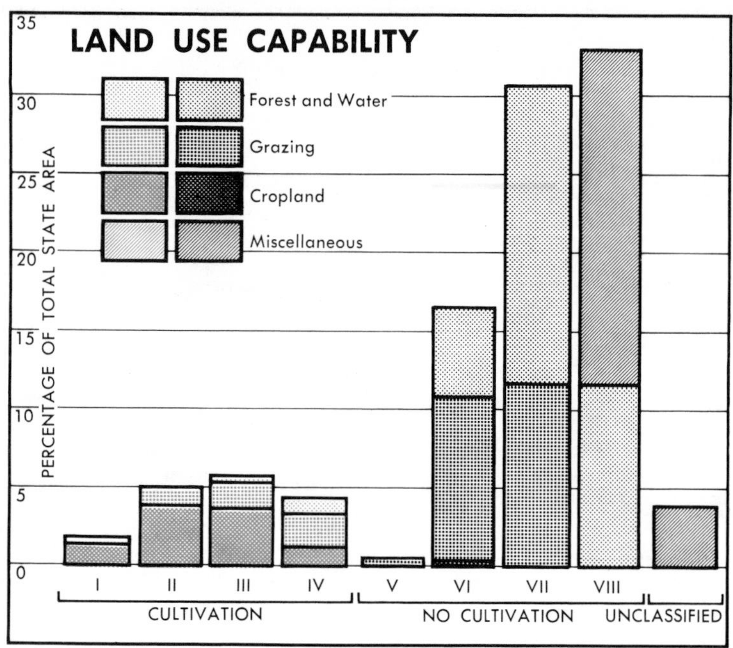

An agricultural area with soils classified according to land-use capability categories.

SOIL

California soils exhibit tremendous variety in their principal characteristics and capabilities. Some of the best and some of the poorest soils found anywhere in the country are present here. Our current pattern of land use emphasizes this point.

Parent material, climate, vegetation and topography are factors which operate to produce differences in soil types from one part of the state to another. Soil depth, permeability, water-holding capacity and nutrient-supplying capacity represent the most important soil properties influencing plant growth.

Soils are classified as *Residual* soils if they are found in the places where they were formed or as *Transported* soils if they have been carried into valley and slope locations by wind or water. *Residual* soils cover well over half the area of California and represent the soils on which most of our forests grow. They vary in their capability to support plant growth according to the nutrients which they possess and on their depth to bed rock. Thin and rocky *Residual* soils are known as *Lithosols.*

Transported soils represent the principal agricultural soils of California and are subdivided into three sub-types: the *Valley, Valley Basin* and *Terrace* soils. *Valley Basin* soils are found in the lowest, most poorly drained parts of the valleys. Factors limiting their use are flooding, drainage problems and clay texture.

The *Valley* soils represent most of the good farm lands of the state. Soils in this category are formed on alluvial flood plains and alluvial fans.

Terrace soils are found in limited quantities along the coast between the mountains and the sea and along streams where older valley flood plains remain in place as gently-sloping aprons of cultivable land. Most areas such as these are too small to be shown on a map. In larger valleys *Terrace* soils are found along the edges of the valleys above the *Valley* soils and consist of older alluvial fan or valley fill deposits which have been exposed to the climatic elements for a considerable length of time.

All of these soil types may be further subdivided on the basis of the climate and vegetation which exist in the area in which they form. Thus, *Residual* soils formed in the humid forests of northwestern California are entirely different from *Residual* soils formed in desert mountains.

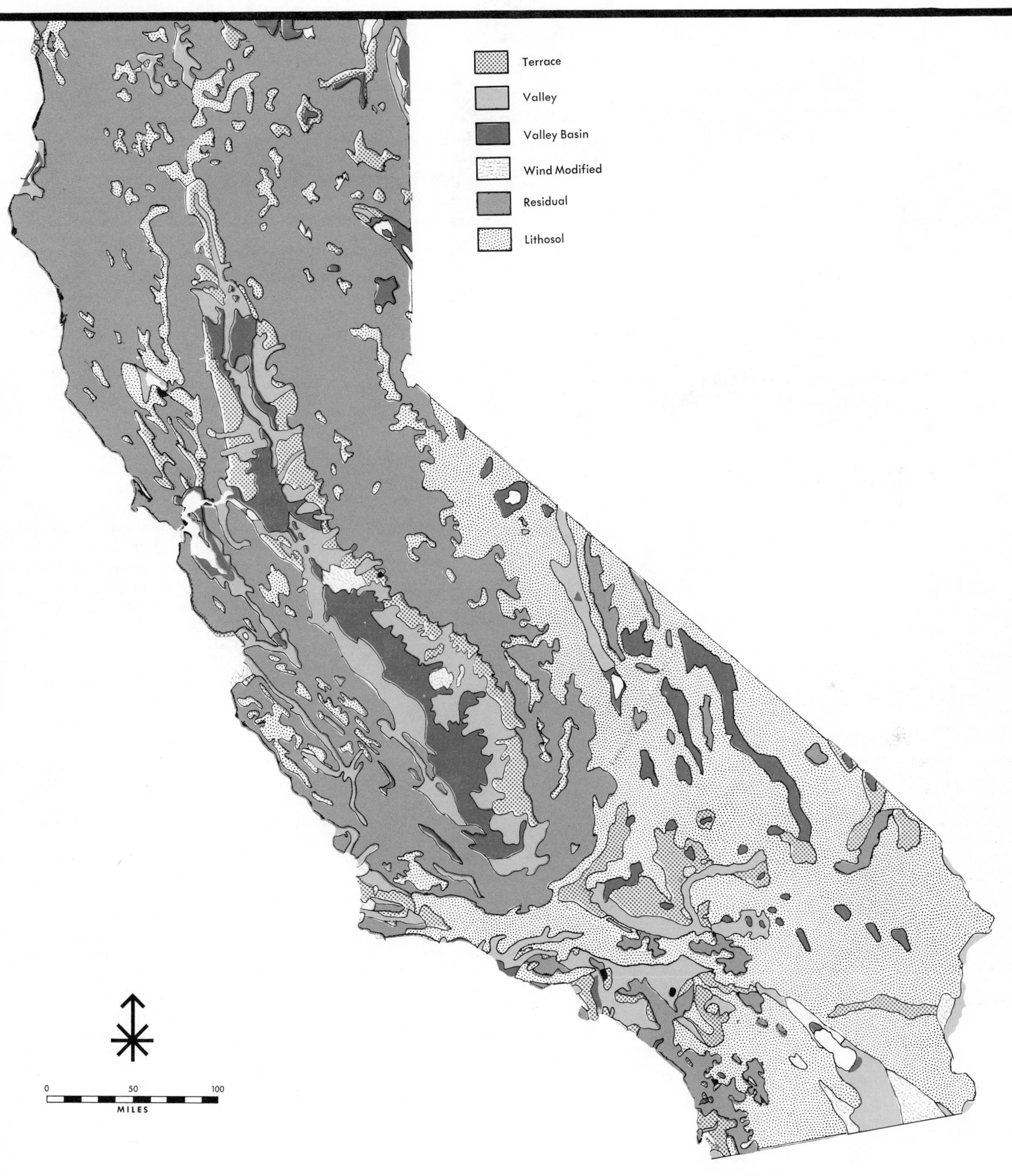

Source: Modified from Storie and Weir, *Generalized Soil Map of California*, 1951.

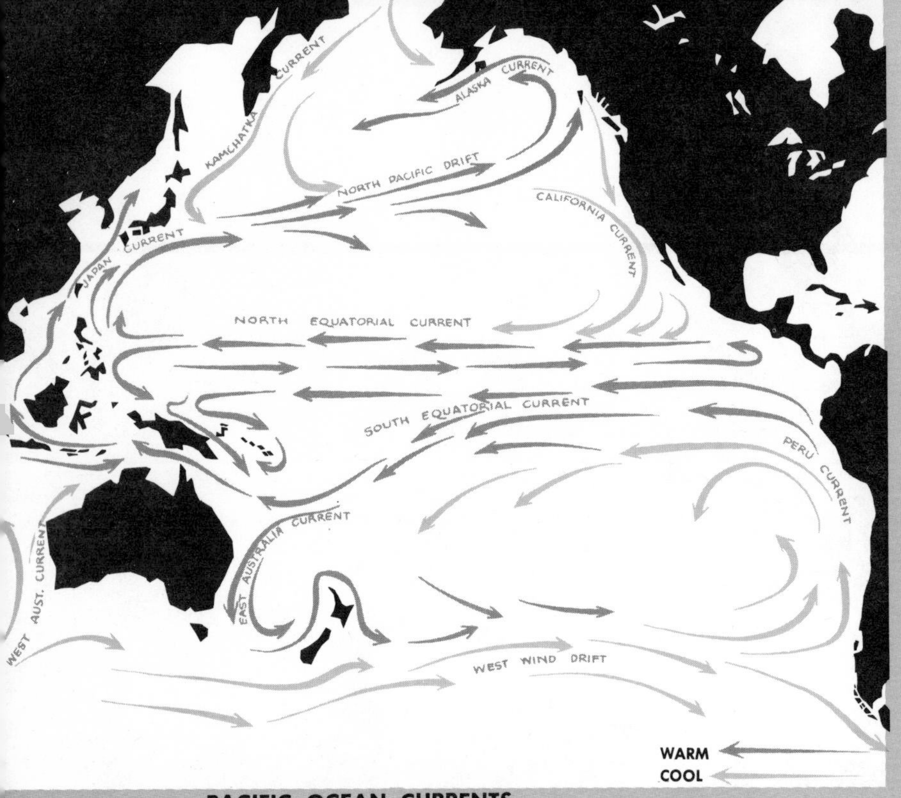

PACIFIC OCEAN CURRENTS

OCEAN TEMPERATURES

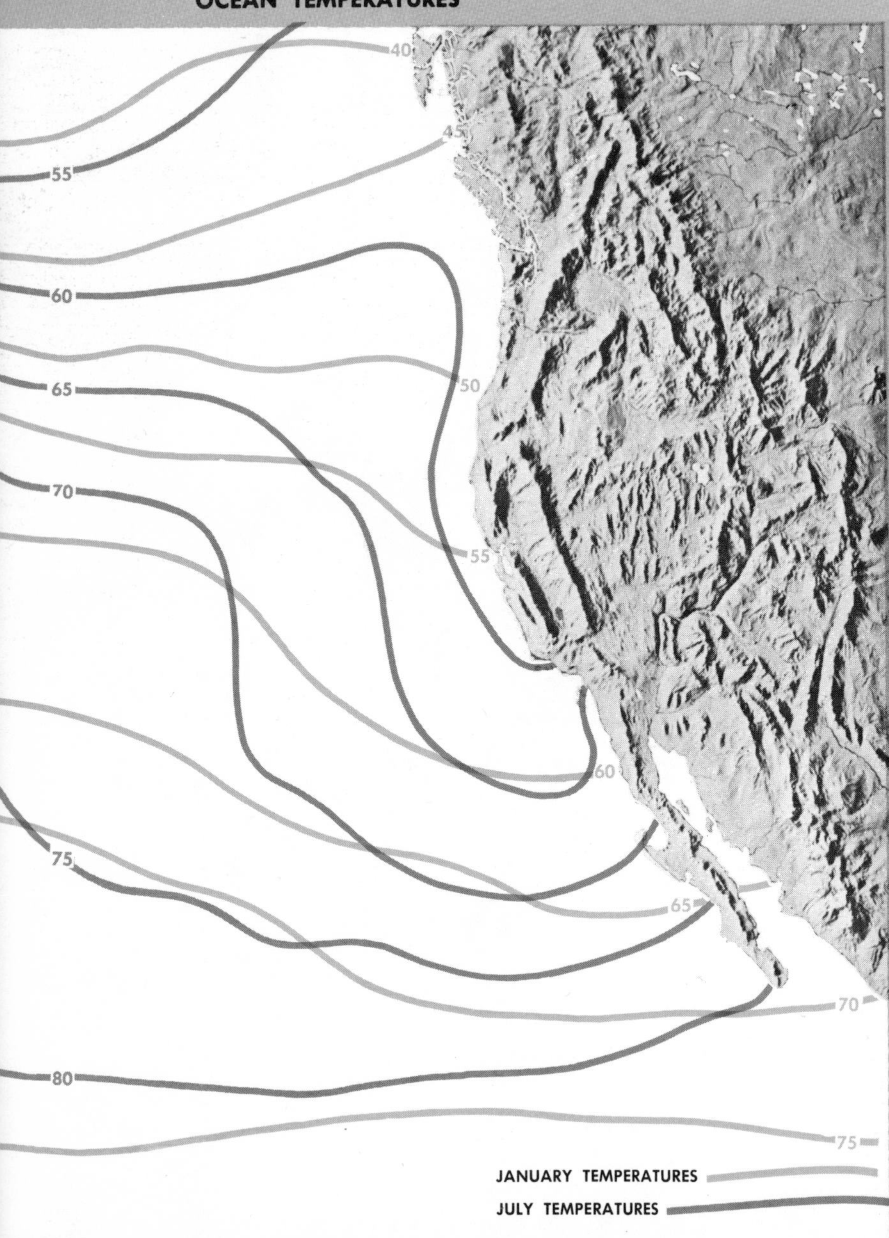

OCEAN CURRENTS

Offshore of the Pacific Coast of the United States flows the California Current, a sluggish meandering stream of water about 400 miles wide. It is a part of the great clockwise circulation of the North Pacific Ocean. At high latitudes the waters move eastward under the influence of strong westerly winds. Near the coast of North America they divide into two branches; the smaller one turns northward into the Gulf of Alaska while the larger part turns southeastward to become the California current.

Between the California Current and the coast lie a number of countercurrents and eddies which are constantly changing with the seasons. A deep counter-current, below 600 feet, flows to the northwest along the coast from Baja California to a point beyond Cape Mendocino. It carries warmer and more saline water great distances northward along the coast. When the north winds are weak or absent in late fall the countercurrent forms at the surface on the inshore side of the main California current and extends from the tip of Baja California to north of Point Conception. It is known as the Davidson current.

The most important force moving the surface of the ocean is the wind. During the summer there are strong northerly winds along the California coast, and the current system is well developed. As the winter season approaches the Pacific High becomes weaker and shifts southward. The northerly winds become less effective and the California Current less well developed.

Along the coast of California winds from the north and northwest are of additional significance. As a consequence of these winds the surface waters are carried offshore and are replaced by waters from below. This process is known as upwelling and is significant for several reasons. It brings extremely cold water to the surface just off the California coastline. This water is richer in nutrient material than the water which has been displaced.

Because upwelling is associated with winds from the north and northwest, it is most marked when these winds are strongest. Such winds are strongest off southern and central California in May and June and in July and August off northern California and Oregon. Upwelling is strongest in the spring months in the south while in the north it is strongest in summer.

California's coastal waters represent a resource which is of ever-increasing importance. In addition to the aesthetic and recreational values associated with the seashore, there are vast supplies of marine fish life available for food. The sea may one day provide our urban centers with water supplies obtained by desalinization of ocean water and may at the same time provide for the mineral needs of our growing population.

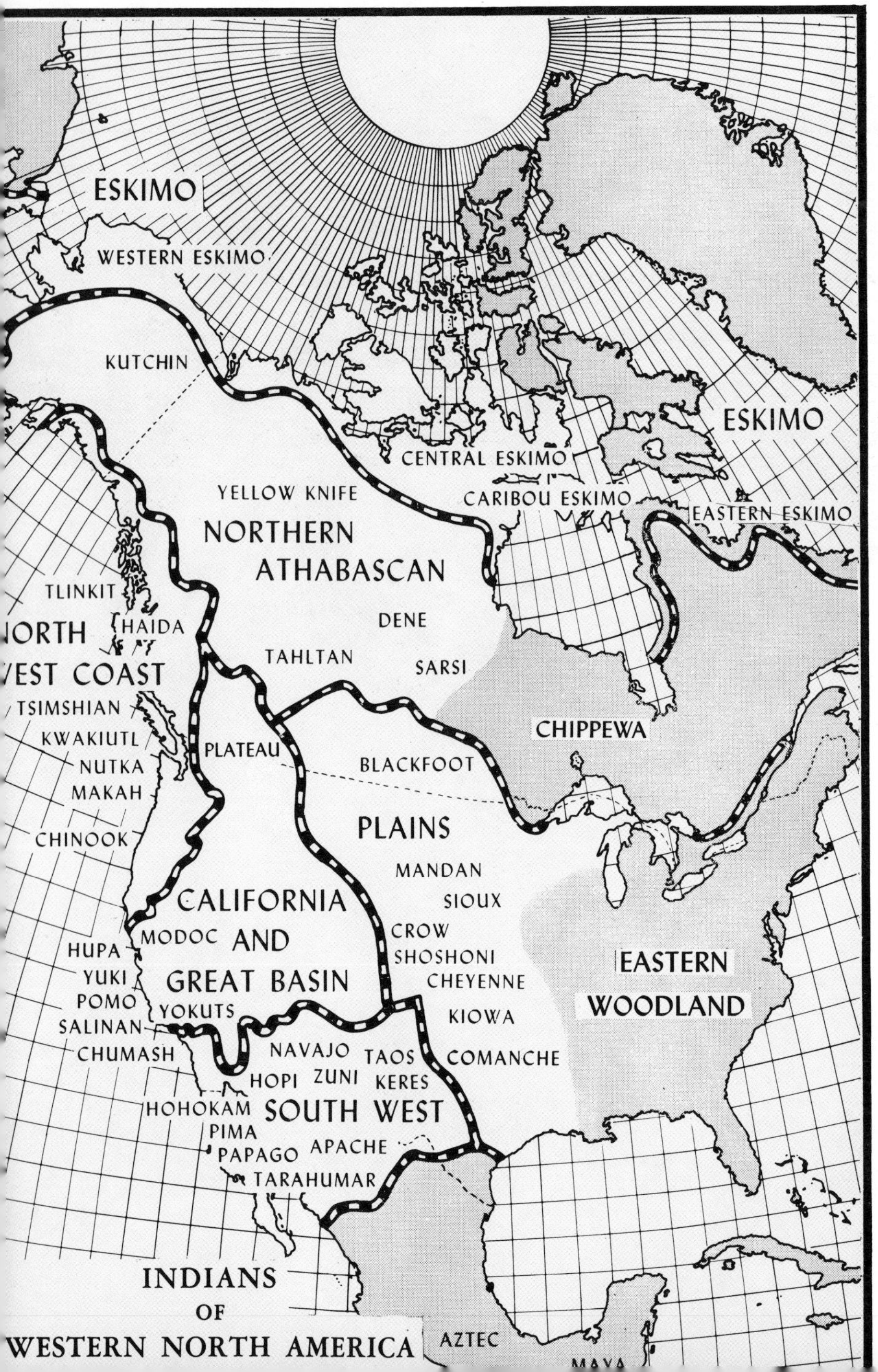

24 INDIANS

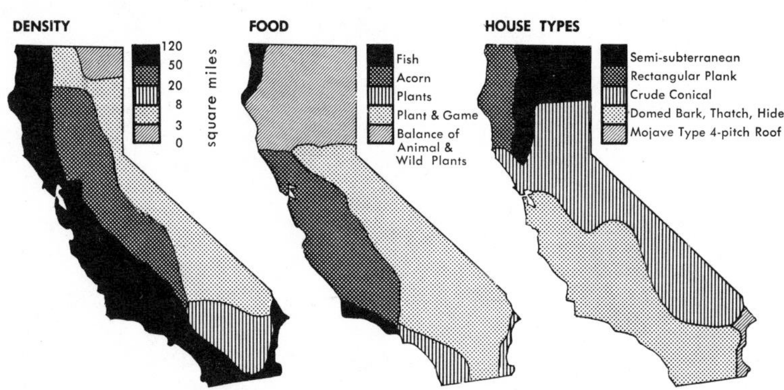

INDIANS

The American Southwest contained a greater variety and a greater number of Indians than any other comparable area in North America. As the Indians came from Asia and filtered southward into the mountainous West, they settled in valleys where water was readily available.

Most of these early settlers were hunters and seed gatherers. This meant deer for those who lived near the mountains, fish for those near streams and rabbits and other small wild animals for the hunters over much of the area. Nets and bows and arrows were used in both fishing and hunting. However, the principal sources of food for these primitive groups were the plants which provided seeds and nuts which could be ground into flour and baked into bread or boiled into a thick soup.

Shelter over much of the area was provided by a structure of poles covered with bark, brush or branches of trees. In northwestern California plank houses with gabled roofs were built out of the redwoods and fir trees of that area.

Most of the villages were relatively small and ranged in size from those with a few related families in the southern mountain areas to those of 1,500 to 2,000 people along the salmon-rich streams of the northwest coast. In all, it has been estimated that there were between 133,000 and 300,000 Indians in California at the time of its discovery by the Spanish. Great numbers of them died as a result of exposure to European diseases and as a result of neglect and mistreatment at the hands of both Mexicans and the Americans. About one-fifth of the Indians now living in California reside on reservations.

ESKIMO
WESTERN ESKIMO
KUTCHIN
ESKIMO
CENTRAL ESKIMO
YELLOW KNIFE
CARIBOU ESKIMO
EASTERN ESKIMO
NORTHERN ATHABASCAN
TLINKIT
HAIDA
DENE
ORTH
TAHLTAN
SARSI
EST COAST
TSIMSHIAN
CHIPPEWA
KWAKIUTL
PLATEAU
NUTKA
BLACKFOOT
MAKAH
PLAINS
CHINOOK
MANDAN
SIOUX
CALIFORNIA AND GREAT BASIN
MODOC
CROW
HUPA
SHOSHONI
EASTERN WOODLAND
YUKI
CHEYENNE
POMO
YOKUTS
SALINAN
KIOWA
CHUMASH
NAVAJO
TAOS
COMANCHE
HOPI
ZUNI
KERES
HOHOKAM
SOUTH WEST
PIMA
PAPAGO
APACHE
TARAHUMAR
AZTEC
MAYA

INDIANS OF WESTERN NORTH AMERICA

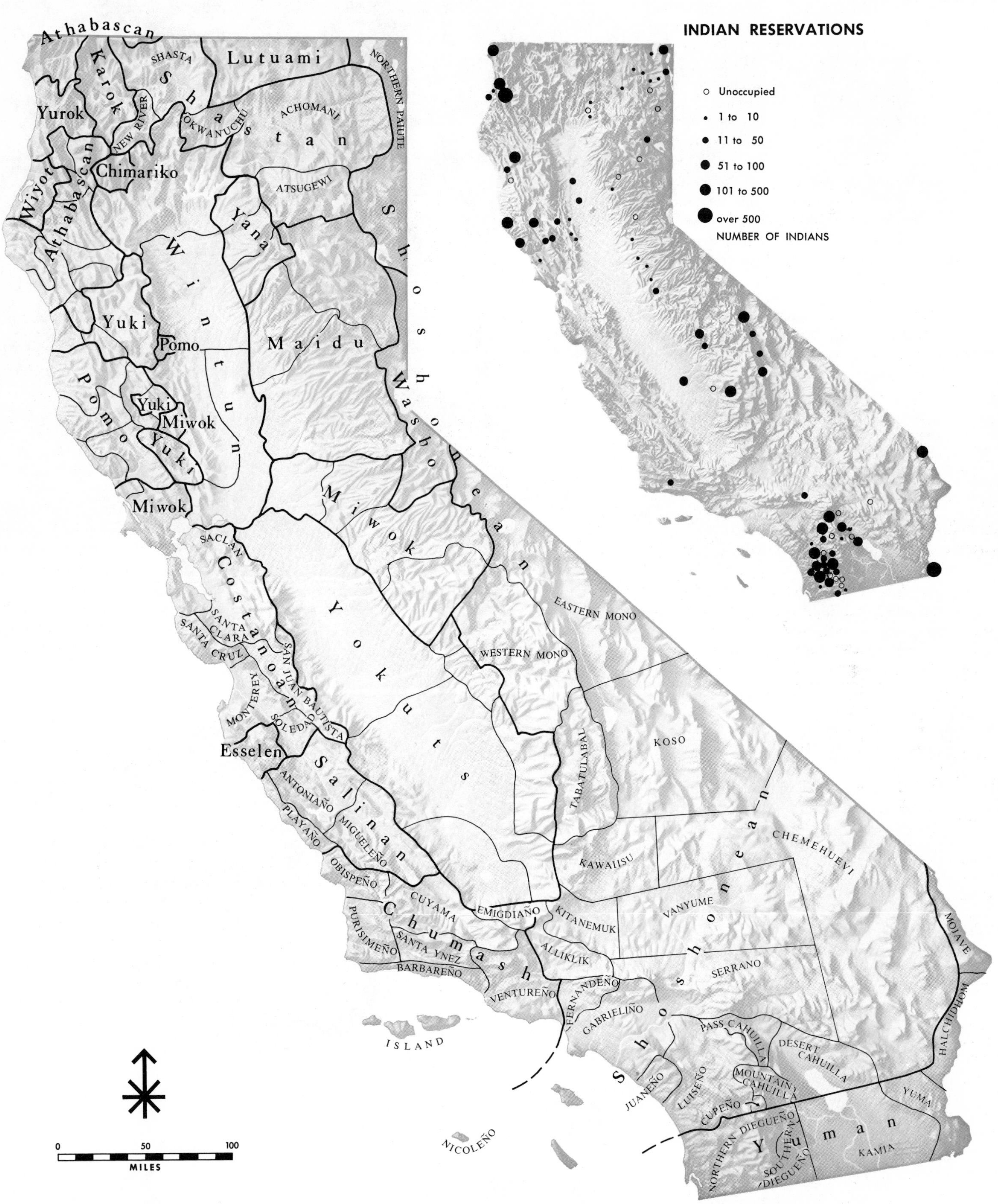
INDIAN RESERVATIONS
Unoccupied
1 to 10
11 to 50
51 to 100
101 to 500
over 500
NUMBER OF INDIANS
Athabascan
Lutuami
SHASTA
Karok
Yurok
Shastan
NORTHERN PAIUTE
ACHOMAWI
OKWANUCHU
NEW RIVER
Chimariko
Wiyot
Athabascan
ATSUGEWI
Yana
Wintun
Shoshonean
Yuki
Pomo
Maidu
Washo
Pomo
Yuki
Miwok
Yuki
Miwok
Miwok
SACLAN
Costanoan
Yokuts
SANTA CLARA
SANTA CRUZ
EASTERN MONO
WESTERN MONO
SAN JUAN BAUTISTA
MONTEREY
SOLEDAD
Esselen
KOSO
Salinan
ANTONIAÑO
PLAYAÑO
MIGUELEÑO
TABATULABAL
CHEMEHUEVI
KAWAIISU
OBISPEÑO
CUYAMA
EMIGDIAÑO
KITANEMUK
VANYUME
PURISIMEÑO
SANTA YNEZ
Chumash
ALLIKLIK
SERRANO
MOJAVE
BARBAREÑO
VENTUREÑO
FERNANDEÑO
GABRIELIÑO
HALCHIDHOM
ISLAND
PASS CAHUILLA
DESERT CAHUILLA
MOUNTAIN CAHUILLA
JUANEÑO
LUISEÑO
CUPEÑO
YUMA
NORTHERN DIEGUEÑO
SOUTHERN DIEGUEÑO
Yuman
KAMIA
NICOLEÑO
0 50 100
MILES

THE SPANISH PERIOD

Columbus' discovery of the New World acted as the spring which catapulted into motion the activities of a whole group of explorers and adventurers who rapidly filled in the details of the map of the world. Within twenty years of the discovery of America, Balboa had reached the Pacific. Much of the Atlantic Coast and of the Gulf Coast of North America had been explored, and Cortés had conquered Mexico. Within thirty years Magellan's crew had circumnavigated the globe and established firmly a number of facts about the size and shape of the world.

Settlement was made first on the island of Espanola (Haiti) in 1493 and for a decade and a half this remained their only Spanish settlement in the New World. Additional settlements were planted on Jamaica, Cuba, Puerto Rico and the Isthmus of Panama between 1508 and 1511. Here the pattern for Spanish conquest and development was perfected.

Operating from their West Indies' bases, the Spaniards searched for gold and for a westward route to the Orient. They began their search in the Caribbean and soon had mapped the eastern shores of the Americas and had penetrated inland in Florida and along other parts of the Gulf Coast area.

By August of 1521 the Spaniards were well established in Mexico City; from there they were able to spread their control over southern Mexico. At first they sought to seize the precious metals and stones which the Indians had accumulated, but they rapidly turned to plantation agriculture with native labor.

Exploration along the Pacific Coast of Mexico was initiated by Cortés who sent ships northward in 1532 and 1533 to explore what was thought to be an island. Following a report of the discovery of a large quantity of pearls near La Paz, Cortés sailed to Baja California in 1535. Few pearls were found and little came of Cortés' attempt to establish a settlement.

Coronado's journey into the interior and Alarcon's voyage up the Colorado were a part of the Spanish quest for additional stores of riches. Lack of any success in finding the seven golden cities of Cibola dulled Spanish interest for further exploration in this direction.

Instead, they returned to the task of establishing a route to the Orient and explored northward along the Pacific Coast of North America.

Christopher Columbus.

COLUMBUS AND MAGELLAN

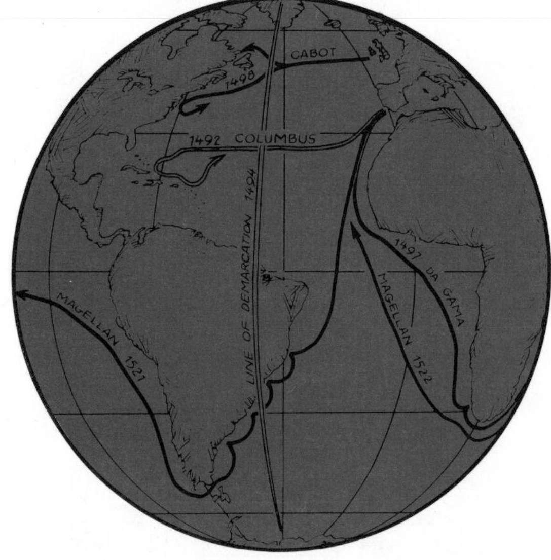

13 Balboa Discovers Pacific Ocean

17-1518 Cordova — Grilalva explores Mexican coast

18-1519 Cortés and army of 600 men conquer Mexico

19 Pineda explores coast of U.S. along Gulf of Mexico

19 Magellan commences voyage around world

Cortez Meets Montezuma.

1520

1522 One of Magellan's vessels under Sebastian del Cano completes voyage around world

1522 Cortés named governor of New Spain

1530

1535 Cortés lands near tip of Baja California

1536 Cabeza de Vaca reaches Culiacan

1539 Ulloa surveys Gulf of California

1540

1540 Coronado begins exploration of Southwest

1540 Alarcón discovers mouth of Colorado River

1541 Bolaños explores coast of Baja California; first used name, "California."

1542 Cabrillo lands at San Diego Bay on Sept. 28.

Juan Rodriguez Cabrillo

1550

ROUTES OF EXPLORATION TO 1550

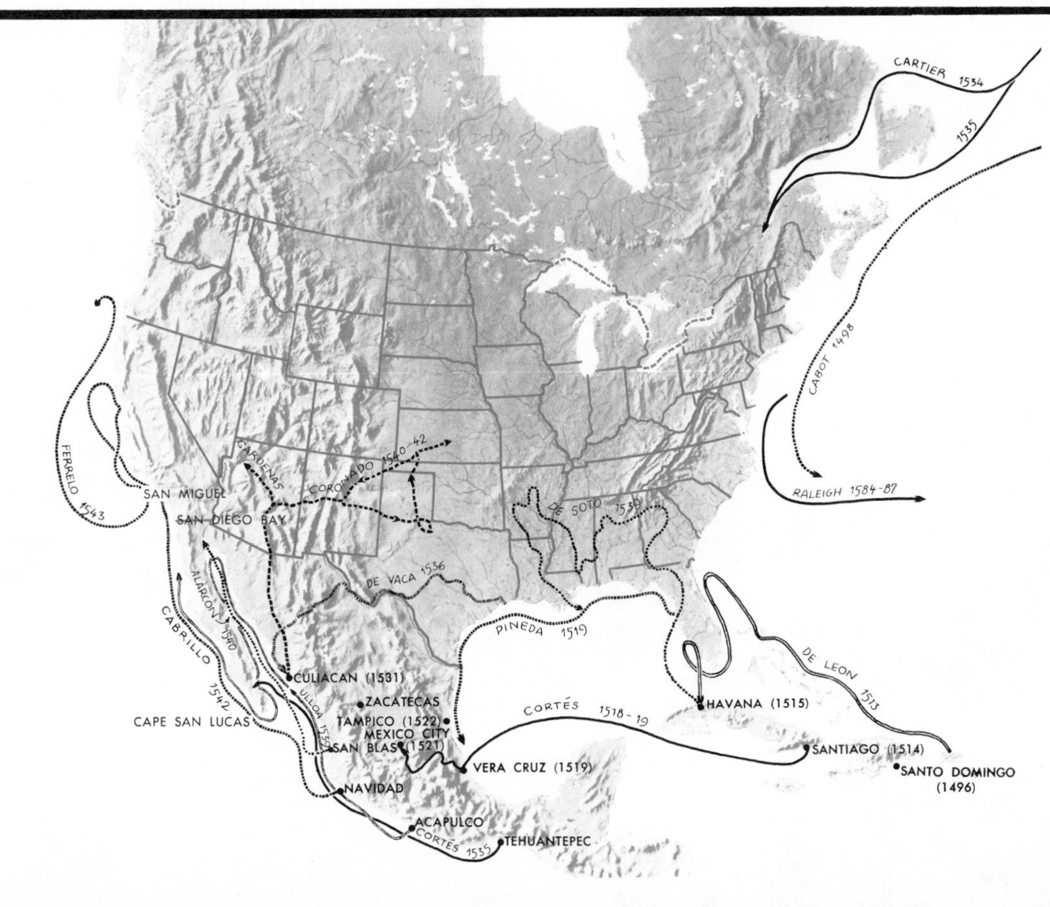

1560

1565 Legazpi-Urdaneta expedition touches California on return to Mexico; establishes Manila Galleon route

1570

1572 Drake sets out on voyage around world

1579 Francis Drake and Golden Hind enter Drake's Bay

1580

1580 Drake completes voyage

1581 Rodriguez explores New Mexico

1582 Espejo explores New Mexic and Arizona

1587 Unamuno explores Californ coast

15

1640

1650

1660

16

On September 28, 1542 Cabrillo discovered San Diego Bay, becoming the first European to reach California. Cabrillo explored the coast northward, landing at several points along the way. Stormy weather brought him back to San Miguel Island where he died on January 3, 1543. Ferrelo, his lieutenant, continued exploration to the vicinity of Cape Mendocino where stormy weather and sickness among his crew forced him to return to Navidad.

Following Cabrillo's discovery of California Spanish interest in further exploration and conquest in this direction slackened. In part this was due to the lack of any reported riches, and in part it was due to the rapid development of the rich silver mines of northern Mexico and the settlement of New Mexico. Additionally, trade with the Phillippine Islands had been established, and the Manila galleons were carrying the riches of the Orient to New Spain. Until almost the end of the Spanish colonial period the

1591 Jesuits undertake mission activity on northwest coast of Mexico

1593 Mexico-Manila trade restricted to one 500-ton vessel

1595 Cermenho explores California coast-discovers Monterey Bay

1598 Spanish Settlement in Rio Grande Valley, New Mexico under Onate

1600

1602 Viscaíno explores and gives names to places along California coast.

1602 Santa Fe established

1605 Oñate reaches delta of the Colorado River

1607 Jamestown Settlement

1610

1615 Iturbe reports that California is an island

1615 Dutch Pirates appear along Pacific Coast

1620

1620 Pilgrims land at Plymouth Rock

1628 Spanish missions established among Hopi in northern Arizona

1630

1673 Fathers Marquette and Joliet explore Mississippi River

1679 La Salle explores Illinois country

1680

1680 Population of New Mexico includes 2400 people of Spanish origin

1680 Revolt in New Mexico and Arizona

1681 Father Hennepin explores upper Mississippi River

1687 Padre Kino enters Pimeria Alta and founds Mission Dolores

1690

1692 Padre Kino founds first mission in Arizona, San Gabriel Guevavi

1696 Founding of Mission Tumacacori in Arizona

1693 Reconquest of New Mexico

1697 Jesuits establish Loreto, first permanent settlement in Baja California

1699 French settlements made in Louisiana and Illinois

1700

1700 Mission San Xavier founded near Tucson

1701 Kino expedition to head of Gulf of California establishes that California is part of the mainland

1710

EXPLORATION AND SETTLEMENT 1550-1710

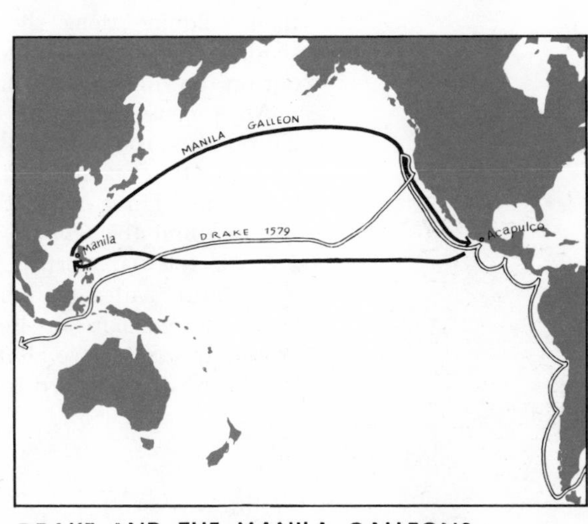

DRAKE AND THE MANILA GALLEONS

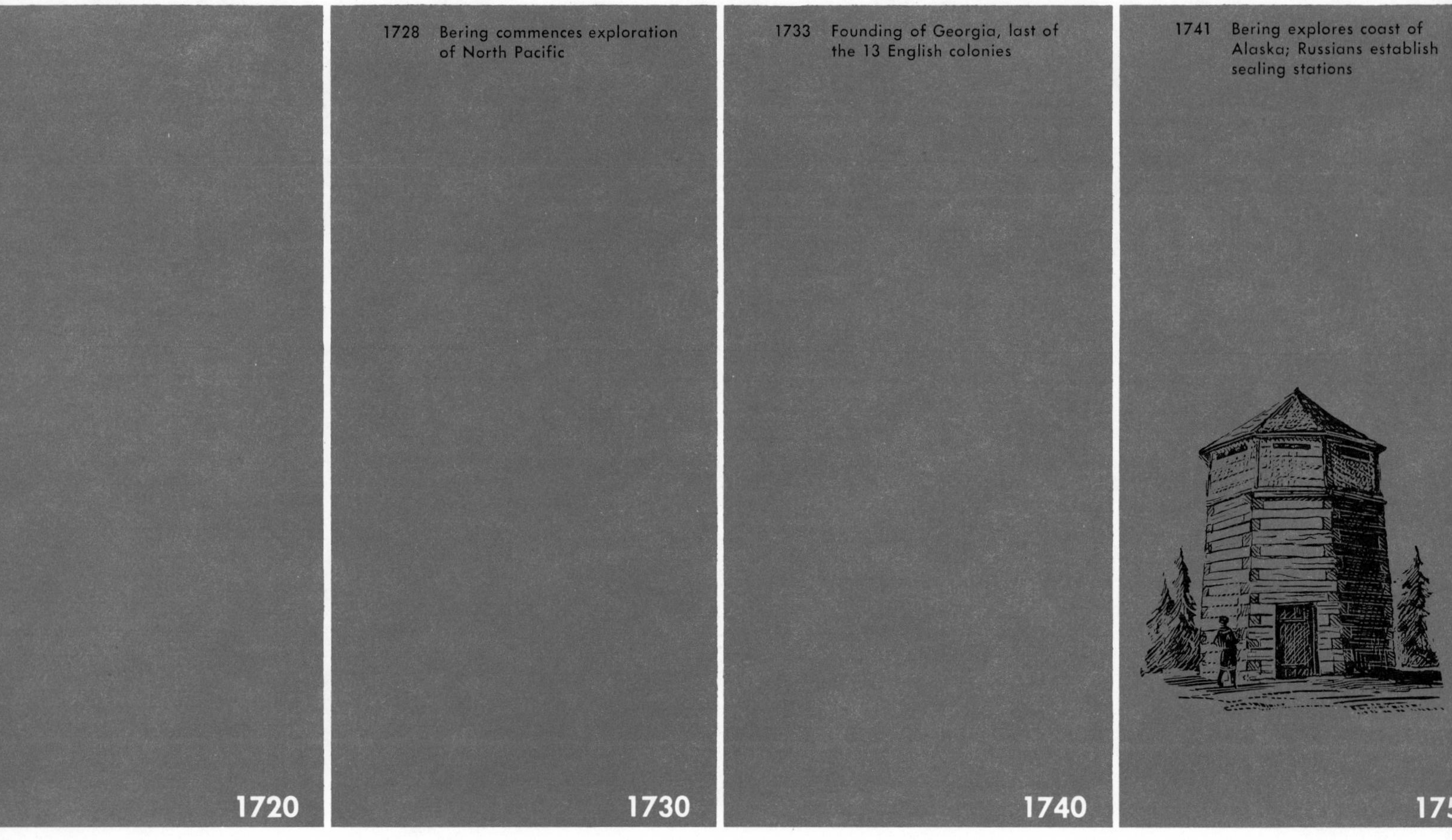

Manila galleon skirted the coast of California. Except for the voyages of Unamuno and Cermeno, landings were infrequent and little knowledge gained.

During this period the French were establishing themselves in the St. Lawrence and Mississippi valleys and the British were establishing their colonies along the eastern seaboard and sending Sir Francis Drake and other adventurers out on the high seas to harass Spanish shipping.

As a consequence of the intrusion of other European nations into the Pacific in the 18th century, the Spaniards finally moved to settle California. Three ships, the *San Carlos,* the *San Antonio* and the *San Jose* were outfitted at La Paz for the trip northward to San Diego. The San Carlos sailed on January 9, 1769 and the San Antonio left on February 15. The third vessel, the San Jose, left on June 16 but was lost with all on board.

The first of two land parties, under command of Rivera y Moncada, left Velicata on March 24 for San Diego. The second land expedition included the commander, Captain Portolá, Father Serra, ten to twelve soldiers, and about forty-four Christian Indians. It started from Loreto on March 9. By July 1, 1769, the four parties were united at San Diego and formal possession of the territory was made. Shortly thereafter, the presidio and mission of San Diego were founded. Portolá and his men continued northward in search of the Bay of Monterey. They eventually found it and are also credited with the discovery of San Francisco Bay.

Almost immediately another presidio was established at Monterey, and additional missions were established at San Gabriel, San Luis Obispo, San Antonio and Monterey. The Spaniards had finally made the move north into modern-day California.

Father Junipero Serra

1752 Founding of Tubac—first presidio in Arizona

1760

1767 Expulsion of Jesuits from New Spain

1769 Captain Cook explores Pacific

1769 Settlement at San Diego. Portolá continues exploration northward to San Francisco Bay

1770

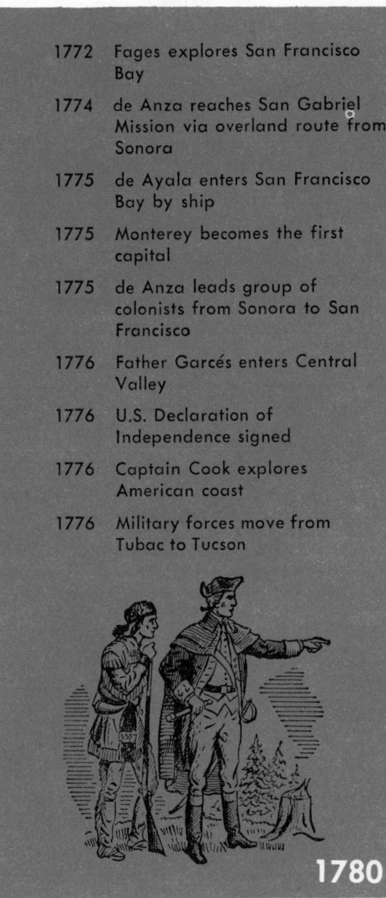

1772 Fages explores San Francisco Bay

1774 de Anza reaches San Gabriel Mission via overland route from Sonora

1775 de Ayala enters San Francisco Bay by ship

1775 Monterey becomes the first capital

1775 de Anza leads group of colonists from Sonora to San Francisco

1776 Father Garcés enters Central Valley

1776 U.S. Declaration of Independence signed

1776 Captain Cook explores American coast

1776 Military forces move from Tubac to Tucson

1780

1781 Pueblo of Los Angeles founded

1781 Yuma Massacre

1784 Father Serra dies; Father Lasuén succeeds him as director of the mission system

1784 Rancho San Pedro, first of large Spanish ranchos granted to Juan José Dominguez

1786 Frenchman, La Pérouse visits California

Mission Santa Barbara

1790

EXPLORATION AND SETTLEMENT TO 1790

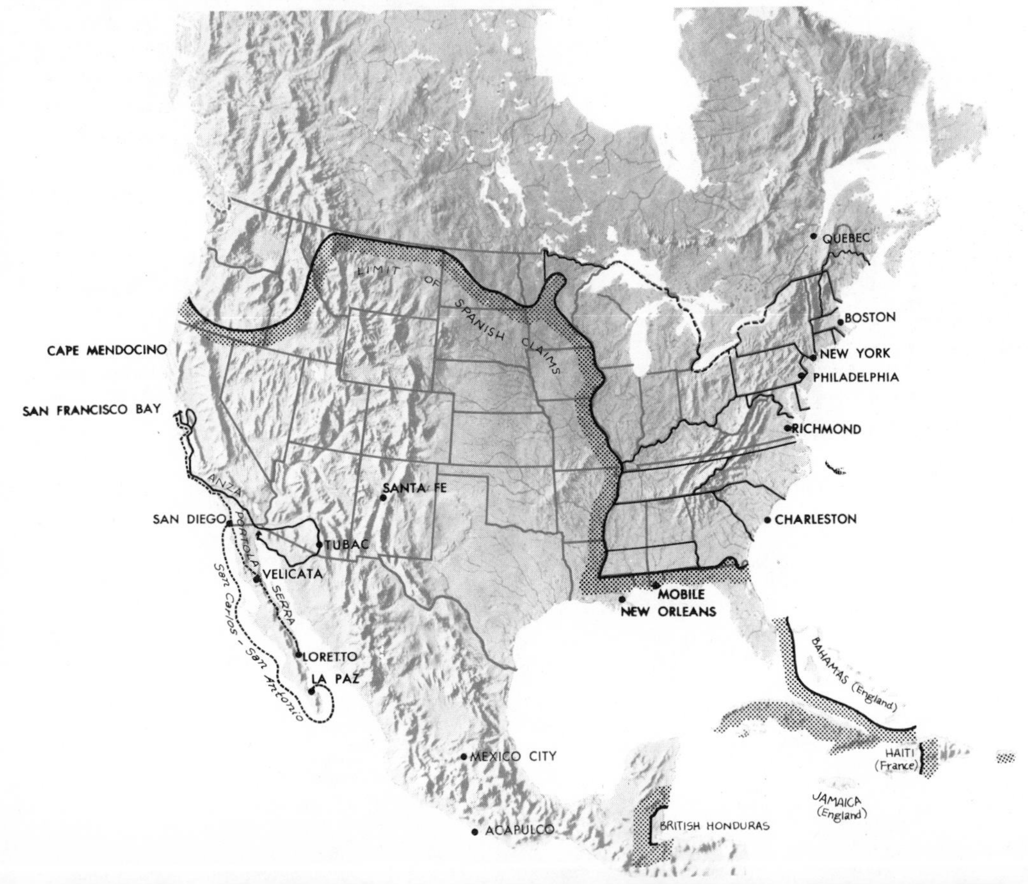

Plan of Mission San Juan Capistrano.

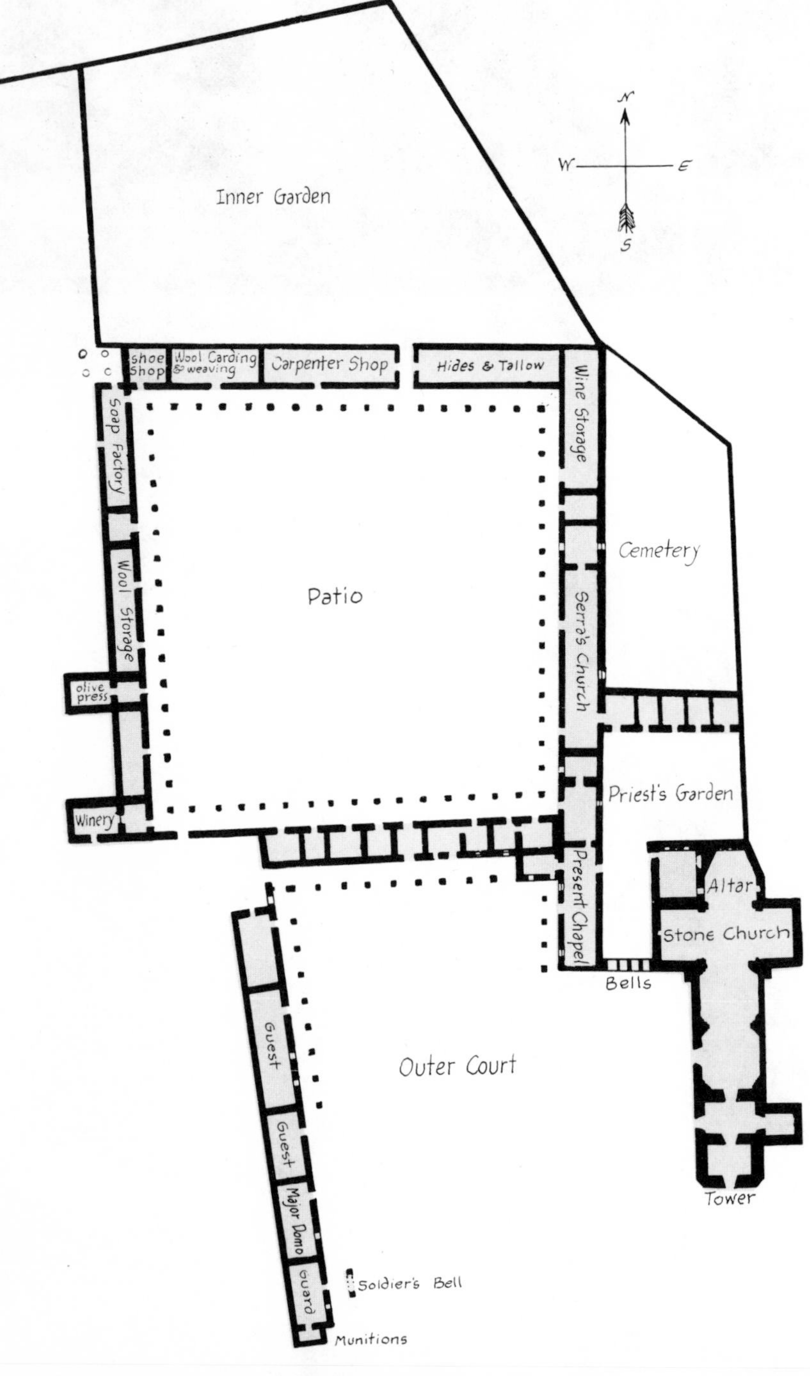

Settlement of California by the Spanish involved a partnership between the religious and the military. During the formative years of the mission system Father Serra labored to establish missions in places where there were adequate numbers of Indians to convert to Christianity and the Spanish way of life. The sites chosen for the missions were near adequate supplies of water for irrigation and for domestic use and in areas where there was suitable land for growing crops and raising livestock. In all, nine missions were built under Father Serra's direction. Father Lasuen, his successor, completed another nine. Presidios were established at strategic sites along the coast to secure the area for Spain and to provide protection for the missions.

Early presidios consisted of an enclosure within which were located a small church, quarters for officers and soldiers, civilian houses, storehouses and workshops. Outside was agricultural land on which food for the soldiers was grown.

The first mission buildings were of crude thatch construction and plastered with mud. Most of the permanent adobe-brick or stone buildings built with Indian labor were erected in the early 1800's.

The first of the civil pueblos was San Jose founded in 1777; next was Los Angeles in 1781; last was Branciforte in 1797. Each pueblo was granted 4 square leagues of land. The central core of the pueblo was the plaza around which were public buildings, homes of the settlers and the parish church. The old Spanish plazas are still to be found in some California towns.

The settlers were either retired soldiers or were recruited in the northern Mexican provinces and brought to the pueblos. They were given a house lot, livestock and implements and an allowance in clothing and supplies for five years. In return, the settlers had to help develop the community by building roads, churches and town buildings and to help till the public lands.

Livestock ranching was a natural outgrowth of the conditions prevailing in this land. There was abundant natural pasturage and a large supply of Indian labor. The small herd of cattle brought to San Diego in 1769 had multiplied rapidly and was readily available to prospective ranchers. The first grant of land was made in 1775 but it was not until 1784 that any large grants were made. At the close of the Spanish period in 1821 only 25 to 30 ranchos were in existence.

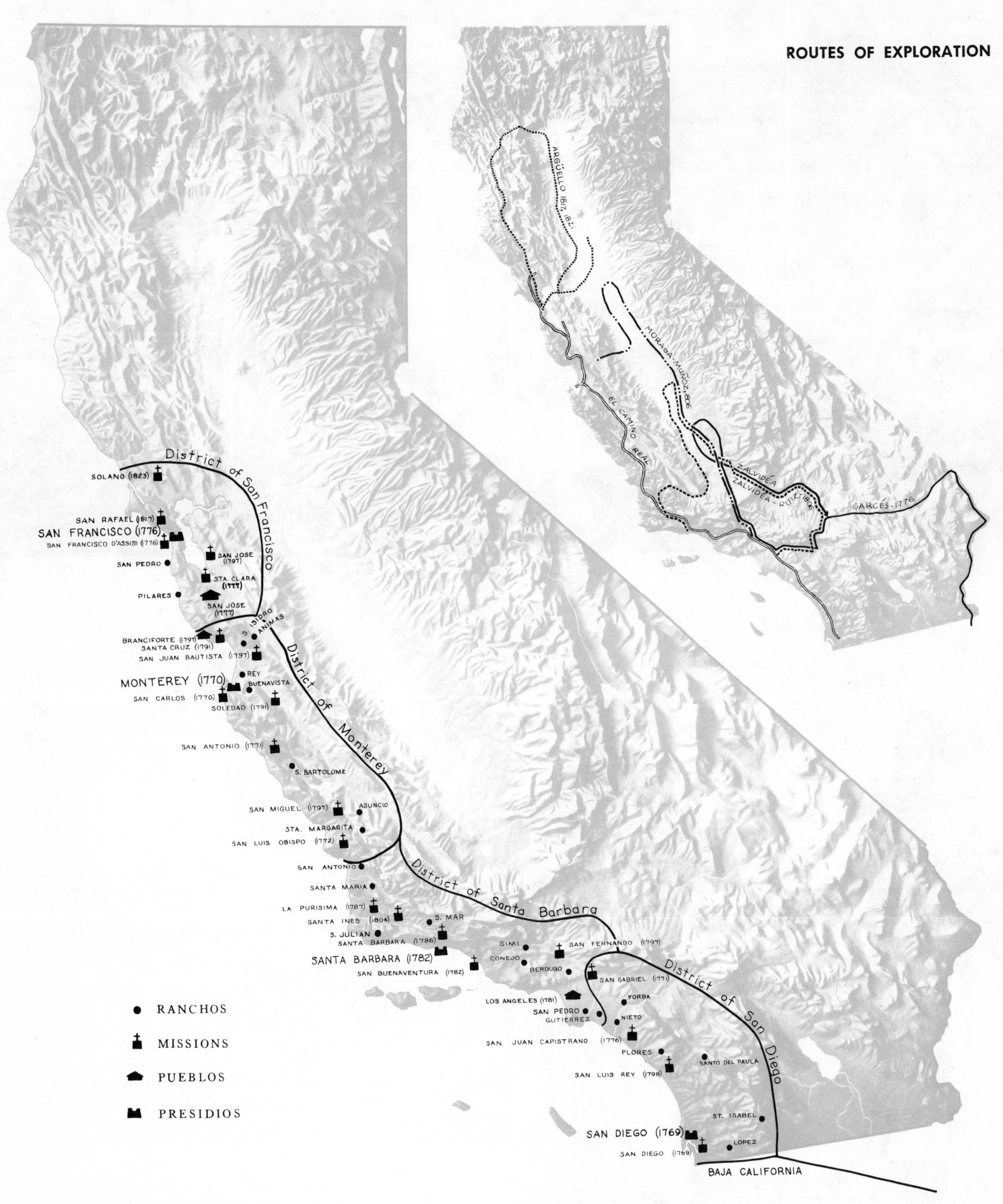
ROUTES OF EXPLORATION
ARGÜELLO 1817, 1821
MORAGA-MUÑOZ 1806
EL CAMINO REAL
ZALVIDEA
ZALVIDEA-RUIZ 1806
GARCÉS-1776
District of San Francisco
District of Monterey
District of Santa Barbara
District of San Diego
SOLANO (1823)
SAN RAFAEL (1817)
SAN FRANCISCO (1776)
SAN FRANCISCO D'ASSISI (1776)
SAN JOSE (1797)
SAN PEDRO
STA. CLARA (1777)
PILARES
SAN JOSE (1777)
S. ISIDRO
ANIMAS
BRANCIFORTE (1797)
SANTA CRUZ (1791)
SAN JUAN BAUTISTA (1797)
REY
BUENAVISTA
MONTEREY (1770)
SAN CARLOS (1770)
SOLEDAD (1791)
SAN ANTONIO (1771)
S. BARTOLOME
SAN MIGUEL (1797)
ASUNCIO
STA. MARGARITA
SAN LUIS OBISPO (1772)
SAN ANTONIO
SANTA MARIA
LA PURISIMA (1787)
SANTA INES (1804)
S. MAR
S. JULIAN
SANTA BARBARA (1786)
SANTA BARBARA (1782)
SAN BUENAVENTURA (1782)
SIMI
CONEJO
SAN FERNANDO (1797)
BERDUGO
SAN GABRIEL (1771)
LOS ANGELES (1781)
YORBA
SAN PEDRO
GUTIERREZ
NIETO
SAN JUAN CAPISTRANO (1776)
FLORES
SANTO DEL PAULA
SAN LUIS REY (1798)
ST. ISABEL
SAN DIEGO (1769)
SAN DIEGO (1769)
LOPEZ
BAJA CALIFORNIA
RANCHOS
MISSIONS
PUEBLOS
PRESIDIOS

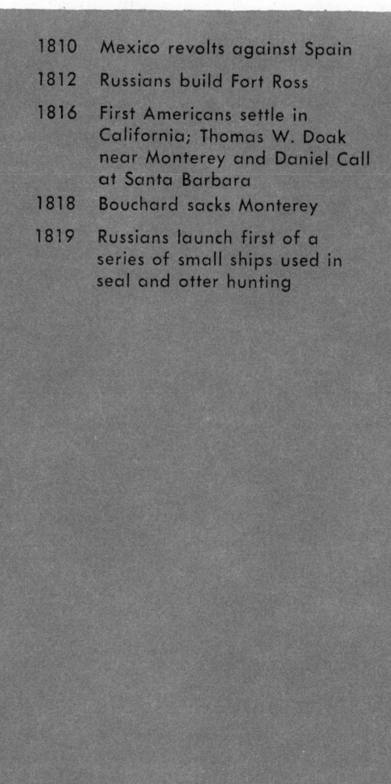

Mountain Men.

THE MEXICAN PERIOD

The revolt of Spanish colonies in the New World followed rather closely upon the heels of the successful revolt of the English colonies along the Atlantic seaboard. The war of revolution in Mexico lasted for over ten years before independence was achieved in 1821. The transfer of authority from Spanish to Mexican officials in California occurred early in 1822.

Relations between the Franciscans and the Mexican government deteriorated very rapidly and ended in the secularization of the missions in the period between 1834 and 1836 when the last mission, Santa Clara, passed out of existence. The missions became parish churches in which the Franciscans performed the religious duties until they were replaced by other priests. There were no specific plans for the disposal of the mission lands except that attempts were made to provide for the welfare of the Indians by allocating some of the land and livestock to them. However, much of the mission property passed into the hands of nearby ranchers, and the land was sold or given away by the administrators. The buildings decayed from neglect, and the orchards and fields of the missions withered away. The Spanish mission system had been destroyed.

Even before the Mexican era began, foreigners had visited California ports. Scientific expeditions sent out by various European nations contributed to knowledge about the Pacific Coast. Otter and seal hunters cruised coastal waters in search of valuable fur-bearing animals. English and Yankee traders on their way to China stopped to take on cargos of furs to trade for the riches of the Orient. Russians, who had initiated the fur trade in Alaska, built a post at Fort Ross in 1812 and gathered a rich harvest along the California coast before leaving in 1841. Whalers careened their ships at Sausalito and Monterey while cattle hide and tallow merchants plied the California coast in search of the raw material for the shoe factories of Massachusetts and England. In 1826 the first of the "mountain men," Jedediah Smith, came to California in search of beaver and on his way back to the Great Salt Lake made the first recorded crossing of the Sierra Nevada. He returned almost immediately to California for more beaver pelts. Another group of trappers which included James Ohio Pattie came down the Gila River to the Colorado in 1827 and reached San Diego only to be thrown into jail. Other trappers working out of Fort Vancouver, St. Louis, Taos and Santa Fe found their way to the beaver-rich streams of the Sierra Nevada.

A number of these early arrivals married native Californians and settled down to become Mexican citizens. Many acquired ranchos and became significant figures in California history. Among these, Sutter and John Marsh became well-known to Americans crossing the continent to California.

Fort Ross — 1828.

ROUTES OF EXPLORATION AND SETTLED AREAS

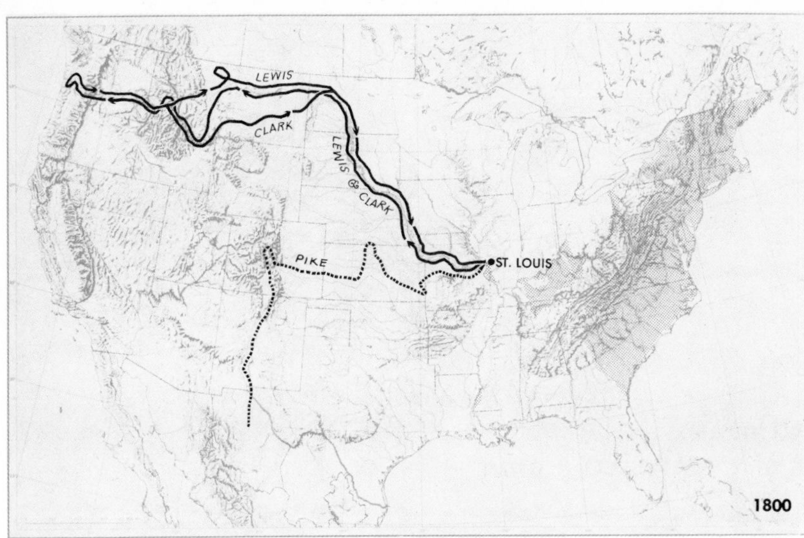

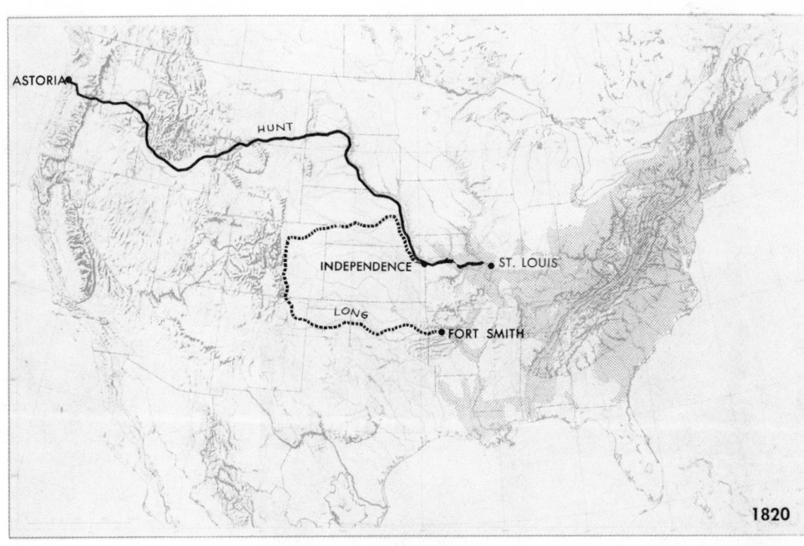

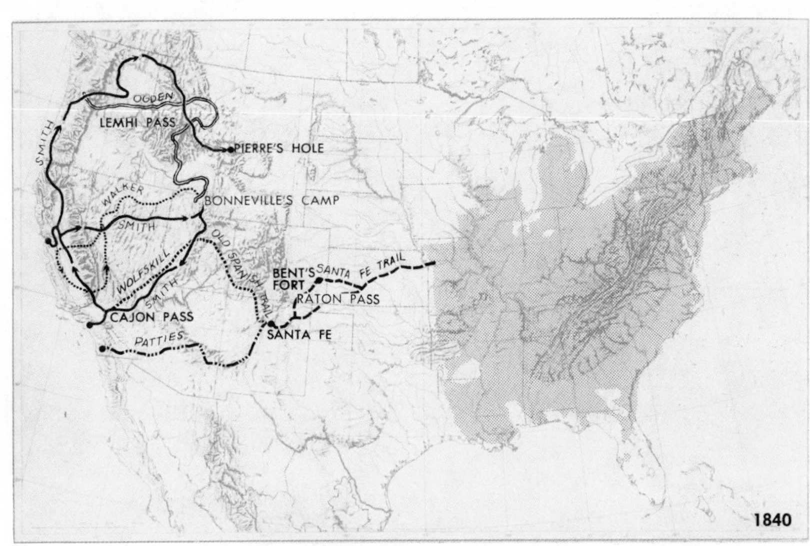

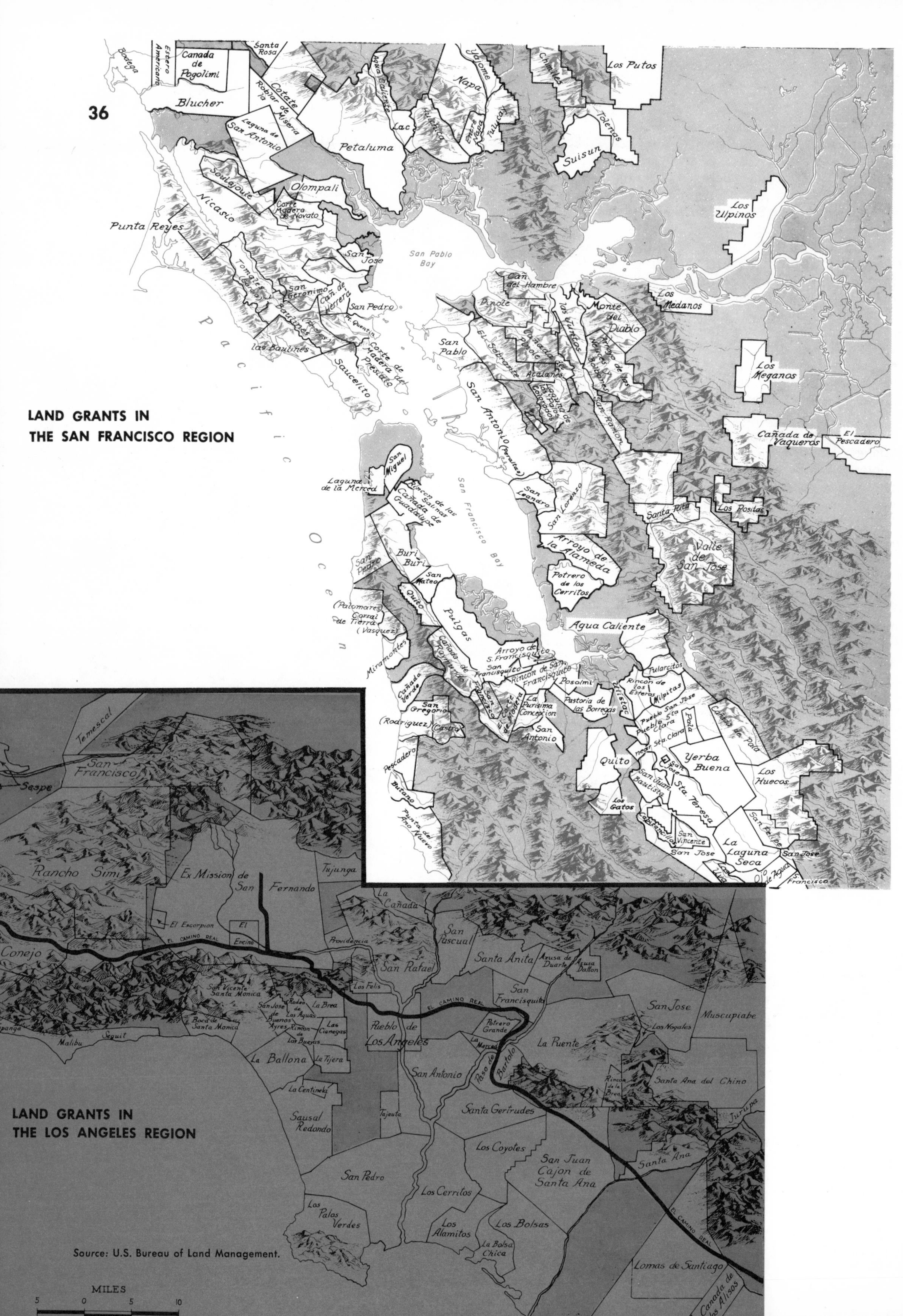
LAND GRANTS IN THE SAN FRANCISCO REGION
Bodega
Estero Americano
Canada de Pogolimi
Santa Rosa
Blucher
Cotate
Roblar de la Miseria
Laguna de San Antonio
Petaluma
Lac
Napa
Los Putos
Suisun
Tolenas
Soulajoule
Olompali
Nicasio
Corte Madera de Novato
Punta Reyes
San Jose
San Pablo Bay
Los Ulpinos
Tomales y Baulinas
San Geronimo
Can. de Herrera
San Pedro
S. Quentin
Las Baulines
Corte de Madera del Presidio
Saucelito
Pacific Ocean
Can. del Hambre
Pinole
Los Medanos
Monte del Diablo
San Pablo
El Sobrante
Acalanes
San Antonio (Peraltas)
San Ramon
Los Meganos
Cañada de Vaqueros
El Pescadero
San Miguel
Laguna de la Merced
Rincon de las Salinas
Cañada de Guadalupe
San Leandro
San Lorenzo
Santa Rita
Los Positas
San Francisco Bay
Arroyo de la Alameda
Valle de San Jose
San Pedro
Buri Buri
San Mateo
Potrero de los Cerritos
Quito
(Palomares) Corral de Tierra (Vasquez)
Pulgas
Agua Caliente
Miramontes
Cañada de Raymundo
Arroyo de S. Francisquito
San Francisquito
Rincon de San Francisquito
Tularcitos
Posolmi
Rincon de los Esteros
Milpitas
Cañada Verde
San Gregorio
La Purisima Concepcion
Pastoria de las Borregas
Pueblo San Jose
Pueblo Sta. Clara
(Rodriguez)
Castro
San Antonio
Pala
Cañada de Pala
Pescadero
Quito
Yerba Buena
Los Huecos
Butano
San Juan Bautista
Sta. Teresa
Los Gatos
Punta de Año Nuevo
San Vicente
San Jose
La Laguna Seca
San Felipe
Ojo de Agua
San Francisco
Temescal
San Francisco
Sespe
Rancho Simi
Ex Mission de San Fernando
Tujunga
La Cañada
El Escorpion
El Encino
EL CAMINO REAL
Providencia
San Pascual
Conejo
San Rafael
Santa Anita
Azusa de Duarte
Azusa Dalton
Los Felis
San Vicente Santa Monica
San Francisquito
San Jose de Buenos Ayres
Rodeo de Los Aguas
La Brea
San Jose
Muscupiabe
Topanga
Malibu
Boca de Santa Monica
Seguit
Rincon de Los Bueyes
Las Cienegas
Pueblo de Los Angeles
Potrero Grande
La Merced
Los Nogales
La Puente
La Ballona
La Tijera
San Antonio
Paso de Bartolo
Rincon de la Brea
Santa Ana del Chino
La Centinela
Santa Gertrudes
Tajauta
Jurupa
Sausal Redondo
Los Coyotes
San Juan Cajon de Santa Ana
Santa Ana
San Pedro
Los Cerritos
Los Palos Verdes
Los Alamitos
Los Bolsas
La Bolsa Chica
Lomas de Santiago
Canada de los Alisos
LAND GRANTS IN THE LOS ANGELES REGION
Source: U.S. Bureau of Land Management.
MILES
5
0
5
10

San Buenaventura
Primer Cañon
Arroyo Chico
Fernandez
Boga
Honcut
Donner Pass
Yokaya
New Helvetia
Sanel
Johnson's Rancho
Guenoc
Canada de Capay
German
Del Paso
San Juan
Rio de los Americanos
Ft. Ross
Sutter's Fort
Cosumnes
Arroyo Seco
Sonoma
Ulpinos
Campo de los Frances
Stanislaus
San Francisco
El Pescadero
Arroyo Seco
El Pescadero
Rancho del Puerto
Mariposas
San Jose
Orestimba
Ausaymas y San Felipe
Panoche
Aguilas
Monterey
San Lorenzo
Laguna de Tache
San Lorenzo
Cholame
El Tejon
Cuyama
Castac
Los Alamos y Agua Caliente
La Liebre
San Emidio
Santa Barbara
0 50 100
MILES
Muscupiabe
San Bernardino
Los Angeles
Jurupa
San Jacinto
Santa Rosa
Temecula
Paula
San Jose del Valle
Santa Ysabel
Valle de San Felipe
Cuyamaca
San Diego
Baja, California

1830 Population: 4,256
1831 Revolt forces Gov. Manuel Victoria to resign
1831 Trapper Young arrives in California with 36 men
1833 Mexican Congress decrees secularization of missions
1833 Walker crosses Sierra Nevada and reaches Monterey
1833 Vignes establishes first commercial vineyard
1834 Zamorano brings first printing press to California
1835 Vallejo founds presidio and pueblo at Sonoma
1836 Alvarado and Castro lead revolt
1836 Wolfskill establishes orchard

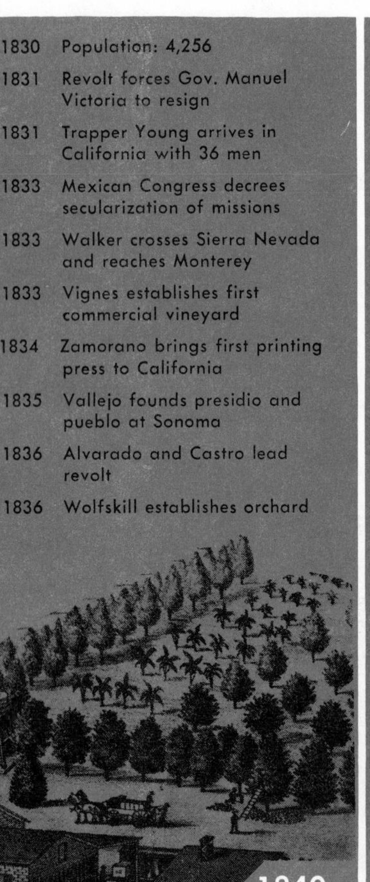

1840 Sutter receives grant, builds fort
1841 Bidwell-Bartleson party arrives
1841 Russians abandon Fort Ross
1842 Gold discovered in Placerita Canyon
1842 Commodore Jones seizes Monterey
1843 Larkin, first U.S. Consul
1844 Fremont arrives at Sutter's Fort
1845 New Almaden mercury mine opened
1846 War with Mexico
Bear Flag revolt
First newspaper, **Californian**, published
1847 Treaty of Cahuenga
1848 Marshall discovers gold; Treaty of Guadalupe Hidalgo
1849 Rush for gold

1850 Statehood granted
1851 Land Commission begins processing claims
1852 Wells Fargo Express established
1853 First Indian reservation at Fort Tejon established
1854 Bret Harte arrives in California
1856 Sacramento becomes the State Capital
1856 First railroad, Sacramento to Folsom; San Francisco vigilante committees form
1857 First overland stage reaches San Diego from San Antonio
1858 Camel Corps arrived in California
1858 Butterfield Stage starts run
1858 First shipment of wheat to New York

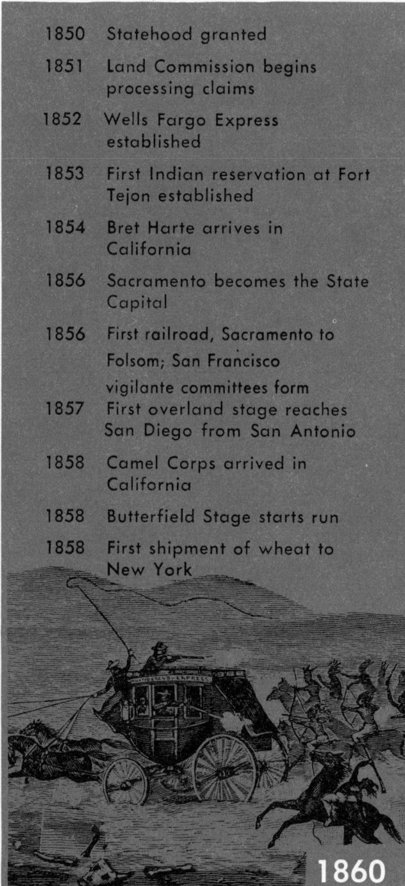

1860 Pony Express delivery arriv
1861 Oil well drilled near Petroli (Humboldt Co.)
1861 Central Pacific RR formed
Transcontinental telegraph completed
1862 Pacific RR Act passed by Congress
1863 Mark Twain in California
Construction of Central Pac RR begins
1868 University of Californa founded
1869 Transcontinental RR comple

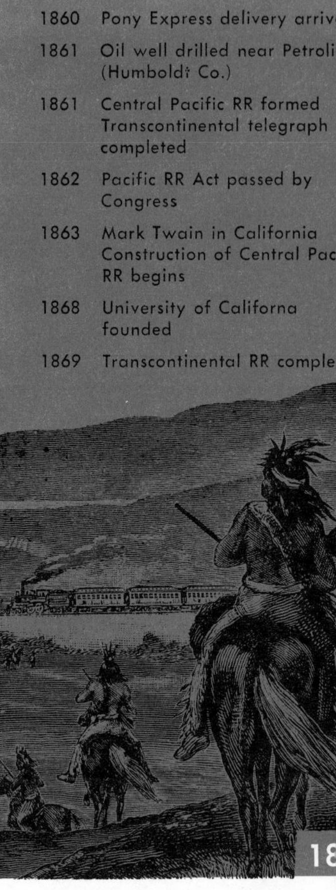

THE AMERICAN CONQUEST

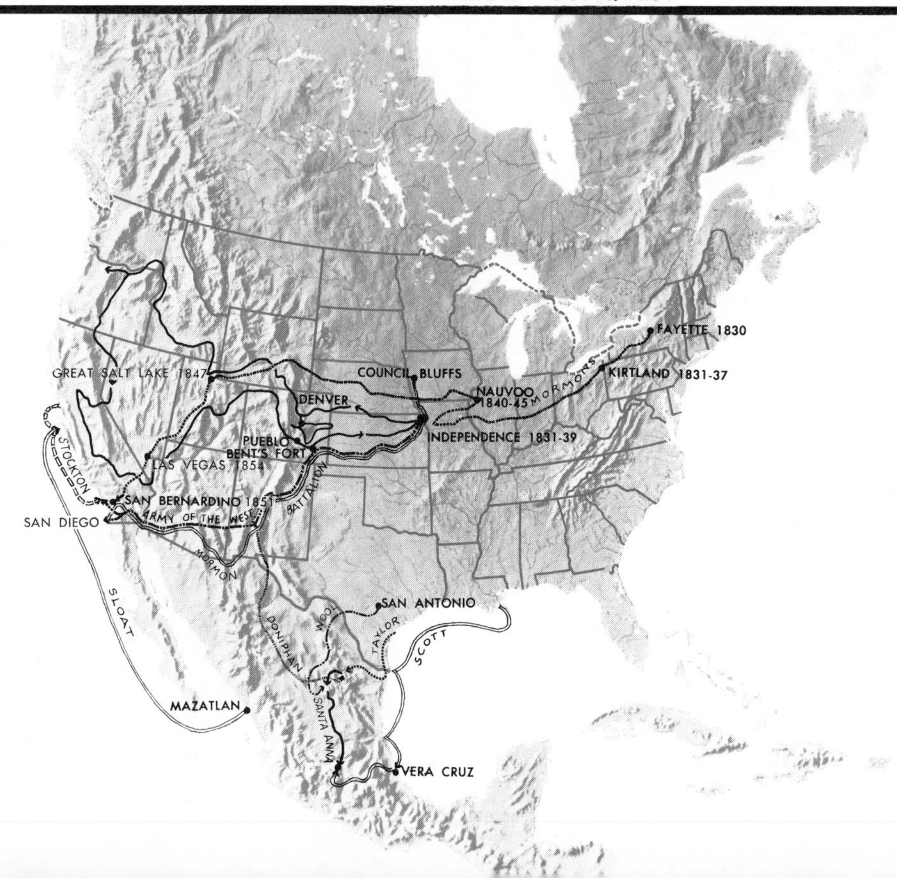

AMERICAN PERIOD

In 1841 a new group of pioneers entered California from the American frontier. The Bidwell-Bartleson party left Sapling Grove near Independence, Missouri in the spring of 1841 along the route of the Oregon Trail. At Soda Springs in southern Idaho they turned southwestward but soon had to abandon their wagons and much of their baggage. Eventually, they reached the Humboldt River and struggled across the Sierra into the Central Valley where they reached the Rancho of Dr. Marsh. A second party of Americans reached southern California the same season having traveled over the Santa Fe and Old Spanish trails in company with the annual group of traders who carried goods between Santa Fe and Los Angeles.

Soon other groups of settlers moved westward along the Overland Trail to settle in the valleys of Oregon and California. Of these the Donner Party became the best known because of their difficulties while snow-bound in the High Sierra.

The Bear Flag Revolt ushered in the period of American control of California. The revolt started when a group of American trappers and settlers managed to capture a herd of horses belonging to General Castro on June 10, 1846. Four days later they took as prisoners Mariano Vallejo and his garrison at Sonoma and carried them off to Sutter's fort for safekeeping. In these activities the Americans received encouragement and direction from Captain John Fremont who had come to California on his third journey of exploration in the West.

News that the United States was already at war with Mexico did not become widely known in California until after July 7, 1846 when Commodore Sloat took possession of Monterey and asserted that, "henceforward California will be a portion of the United States." Within a week the American flag was flying at Yerba Buena (San Francisco), Bodega, Sutter's Fort and Sonoma. This movement of United States military forces to take control of strategic positions in California marked the end of the Bear Flag Republic established by William Ide and his followers.

The war between the United States and Mexico was fought principally in Texas and the Valley of Mexico but there was some activity in California. On July 15 Commodore Sloat was replaced in command by Commodore Robert Stockton who pursued a more vigorous role in

GROWTH OF THE PUBLIC DOMAIN

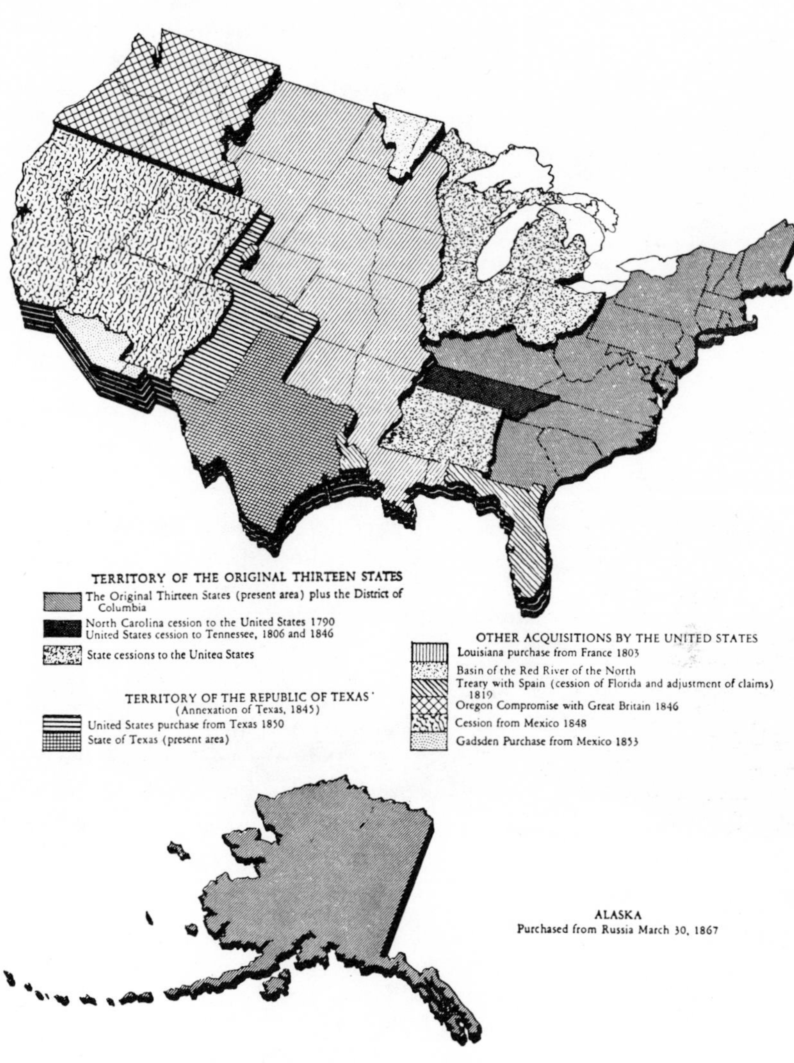

ROUTES OF TRAVEL AND SETTLED AREAS

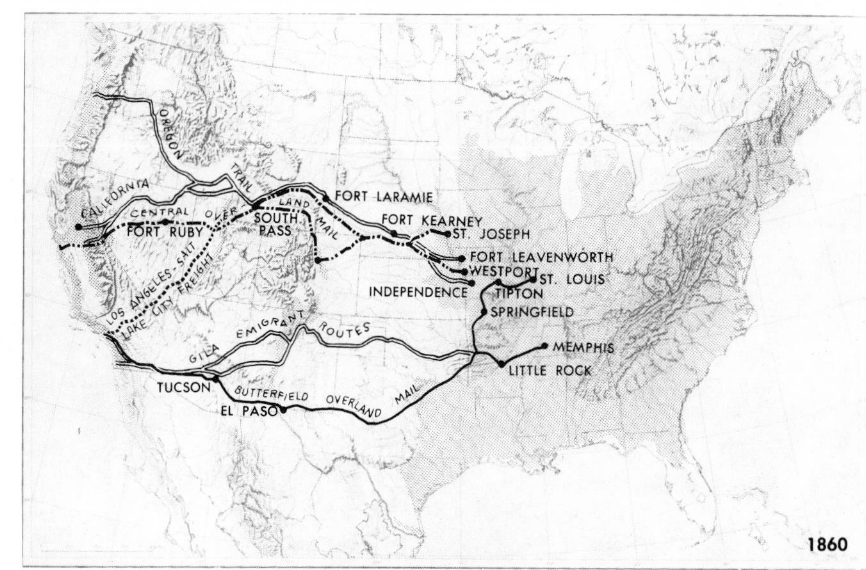

the conquest of California. Fremont, the Bear Flaggers and other American settlers were organized into an army which marched with Stockton's forces over much of California raising the American flag. Their only difficulties arose in southern California where the native Californians recovered from their early state of disorganization and revolted against the Americans. Following a number of minor skirmishes in which a small number of casualties were inflicted on both sides, American troops under Stockton and General Kearny recaptured Los Angeles on January 10, 1847. Three days later Andres Pico, commander of the last armed Mexican force surrendered to Fremont and signed the Treaty of Cahuenga thus ending hostilities in California.

However, the war continued in Baja California and other parts of Mexico for some time after this. The Treaty of Guadalupe Hidalgo was signed on February 2, 1848 formally ending the war and adding the American Southwest to the territory of the United States. Two years later California became the thirty-first state to be admitted to the Union.

GOLD

Discovery of gold by James Marshall at Coloma on the south fork of the American River on January 24, 1848 marked one of the most significant events in the history of our country. It moved the American frontier two thousand miles to the west and brought hundreds of thousands of settlers to California from all parts of the world.

Marshall, in partnership with Sutter, had constructed a sawmill to provide lumber for the growing California population when he plucked a nugget from the river channel. Soon gold was found by other employees, and word of the discovery spread across California.

By May California's towns were deserted by their male inhabitants who had gone to the mines in search of gold. Within months news of California gold had found its way to every part of the globe.

However, only a few thousand miners reached California in 1848; the population increased by about 10,000 during that year. After President Polk gave the gold discovery publicity in his presidential message of December 5, 1848, a tidal wave of humanity commenced to roll toward California.

Those inhabitants of the eastern United States and of Europe who were in a hurry to get started on their way booked passage on ships sailing around Cape Horn. Though there was little danger associated with this voyage, it did require a stout stomach and enough money to pay the fare. Many individuals chose this route, particularly in the first several years of the gold rush.

Another route by way of the Isthmus of Panama involved sea passage to the Panama coast, thence a walk or ride across to the Pacific where the gold seeker boarded another vessel bound for California. Under ideal conditions this represented the quickest way to get to California, but the crossing of the Isthmus presented many problems including malaria.

The greatest number of gold-seekers arrived in California in wagon trains or on horseback after crossing the West over one of the trails established by the trappers and early settlers. Most families on the frontier had a wagon, oxen and horses with which to make the trip; thus, did not need a large amount of capital in order to start their journey. Most came over the Central Overland Trail although some followed the Old Spanish Trail into southern California.

Mining originally involved very simple methods of digging and washing sand and gravel in search of gold flakes and nuggets. Mining spread from the vicinity of Coloma north and south in the foothills of the Sierra and eventually to all parts of California. As the richest deposits were exhausted, more refined techniques for extracting gold from the gravels were developed, and miners began tc dig into the veins from which the placer gold was derived. After 1852 the amount of gold taken from the mines decreased, and miners began to drift off into other occupations and away to other gold fields in Australia, Canada and other parts of the United States.

Gold

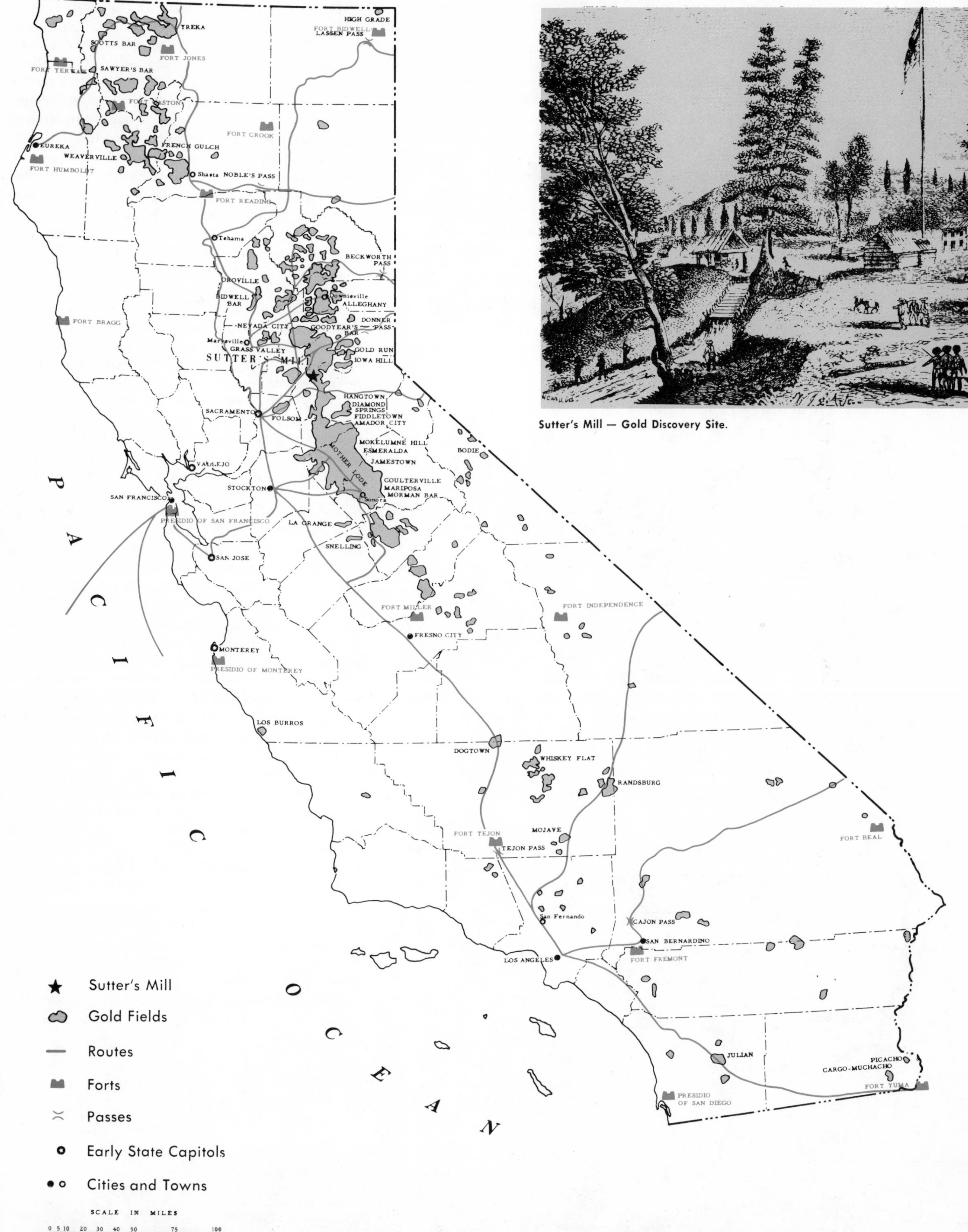

Sutter's Mill — Gold Discovery Site.

Central Pacific Crossing the Sierra Nevada Range.

FEDERAL LAND GRANTS TO RAILROADS

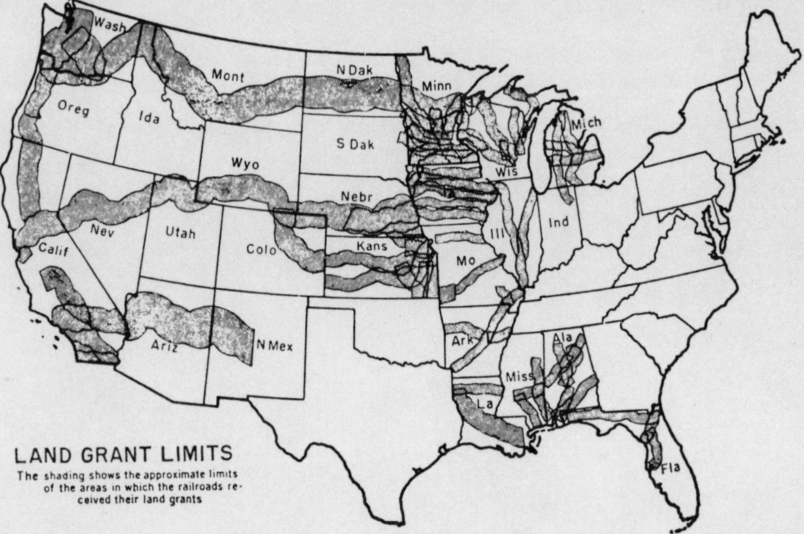

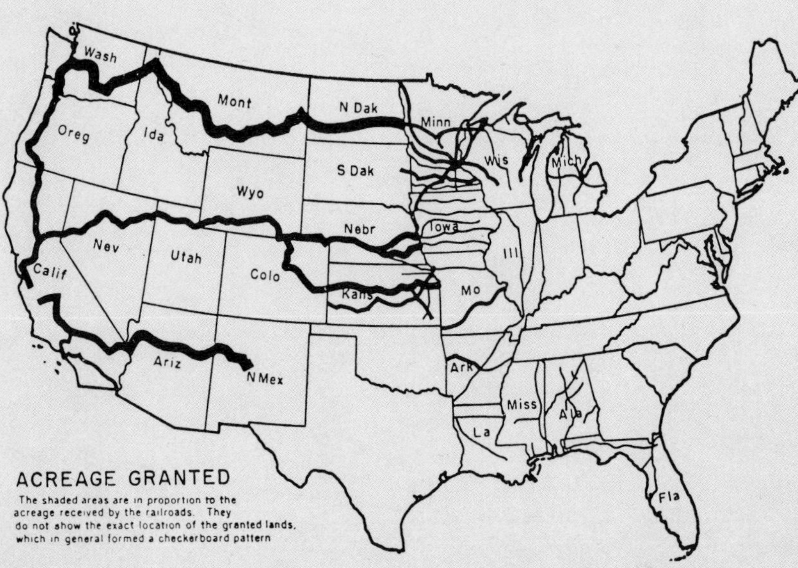

RAILROADS

The next significant event in the settlement of California was the completion of the transcontinental railroad in 1869. Prior to that time and subsequent to the initial years of the Gold Rush, California had been connected to the East by various stage lines, the Pony Express, the transcontinental telegraph, clipper ships sailing around Cape Horn and by steamship via the Isthmus of Panama. However, none of these provided all of the services desired by individuals interested in traveling to California or in shipping goods and mail to and from the West Coast.

In California a number of short railroad lines had been completed in the 1850's. One of them, the Sacramento Valley Railroad which ran from Sacramento to Folsom, was the first segment of the Central Pacific Railroad Company.

Gradually, the idea of a transcontinental railroad to connect California with the East received acceptance, and a number of parties of army engineers surveyed possible routes to the West. In 1861 the outbreak of the Civil War helped decide the question of when and where the railroad should be built. Military and politial considerations dictated the choice. On June 28, 1861 the Central Pacific Railroad was founded in Sacramento by Leland Stanford, President; Collis P. Huntington, Vice President; and Mark Hopkins, Treasurer. These three were shortly joined by Charles Crocker who supervised construction of the road.

The Pacific Railroad Bill was passed by Congress on July 1, 1862 and construction of the railroad commenced in 1863 with the Central Pacific building eastward across the Sierra Nevada, and the Union Pacific building westward from Omaha. Irish immigrants provided the principal labor force for the Union Pacific while Chinese coolies built the Central Pacific. On May 10, 1869 a formal ceremony in which a golden spike was driven marked the joining of East and West. Now it took only seven days to travel from New York to Sacramento.

In the process of building the railroad vast grants of land and cash subsidies had been given to the companies. The Railroad Act of 1862 gave the railroads alternate odd-numbered sections of land for ten miles on each side of the road in addition to a 400 foot wide right-of-way. The 1864 act extended the grant to include lands for twenty miles on each side of the road. Similar grants were made to other railroads in the United States.

After completion of the Central Pacific, the "Big Four" extended their activities into the

Camp Lincoln
Crescent City
Klamath
Yreka
Ft Jones
Mt Shasta
Orleans Bar
Camp Gaston
Camp Anderson
Eureka
Humboldt
Weaverville
Ft Crook
Round Val.
Ft Bidwell
Shasta
Ft Reading
Mt Lassen
Susanville
Honey L.
Camp Grant
Red Bluff
Tehama
Ft Wright
Long Valley
Cahto
Chico
Oroville
Ukiah City
Mt Ripley
Colusa
Marysville
Nevada
Dutch Flat
Donner Pass
Auburn
Placerville
Knoxville
Geyserville
Santa Rosa
Napa
Vacaville
SACRAMENTO
Ebbetts Pass
Petaluma
Benicia
Vallejo
Martinez
Stockton
Pt Reyes
SAN FRANCISCO
Farallones Is
Oakland
Mt Diablo
S. Jose
San Mateo
San Jose
Jackson
Mokelumne
Sonora Pass
Columbus
Sonora
Jacksonville
Sampsonville
Mariposa
Fremonts
Merced R.
SIERRA DEL MONTE DIABLO
Gilroy
Pacheco Pass
Santa Cruz
Monterey Bay
Point Pinos
Monterey
Carmel Pt
Salinas
Pt Sur
San Antonio
San Simeon
Esteros Bay
San Luis Obispo
Pt San Luis
Bay of San Luis
Pt Sal
Oliveras
Sierra de San Rafael
Point Conception
Sta Barbara
Channel of Santa Barbara
San Miguel
Santa Rosa
Santa Cruz
Anacapa
La Panza
Buenavista L.
Kern L.
Tejon Pass
Ft Tejon
San Francisquito Pass
S. Fernando
Los Angeles
Wilmington
Pt Vincent
San Pedro
Anaheim
Sta Catalina
S. Mateo
Santa Barbara
San Nicolas
San Clemente
S. Luis Rey Mis.
Warners
Warners Pass
Sta Felipe
San Diego
Point Loma
Ft Yuma
Valley below the Ocean
San Jacinto Mountains
Coahuila Valley
San Bernardino Mts
St Bernardino Mt.
Halcomb Valley
Cayoto Pass
Camp Cady
Mohave R.
Camp Rock Sp.
Providence Mts
Kingstone Sp
Hernando Sp.
Camp Eldorado
Old Mormon Ft
Dry Lake
Death Valley
Salt L.
Dry L.
Hum-pah-ya-map Pass
Kern R.
Tulare Lake
Fresno City
Woodville
Camp Rabbit
Millerton
Mt Humphreys
Owens R.
Camp Independence
Bend City
Mt Williamson
Owens L.
Camp Fish L.
Silver Peak City
Palmetto
Benton
Aurora
Columbus
Castle Peak
Mono L.
Adobe Meadows
Genoa
Carson
Tahoe L.
Pyramid Peak
VIRGINIA
Ft Churchill
Empire
Pine Nut Range
Walker L.
Ellsworth
Ione
Union
Belmont
Toiyabe City
Hot Creek
Austin
Jacobsville
Toyabe Mountains
Stillwater P.O.
Sink of Carson
Sink of Humboldt
Humboldt L.
Murphy P.O.
Parker
Pyramid L.
Trinity Mts
Unionville P.O.
Star City
Humboldt City
Camp McKee
Smoke Cr. Depot
Eagle L.
Madeline Plains
Camp McGarry
Lassens Road
Antelope Mountains
Black Rock Desert
Santa Rosa Mounts
Camp Winfield Scott
Scottsville P.O.
Camp McDermitt
Disaster Peak
Pueblo City
Centreville
Camp Lyon
Humboldt Pass
Wells Sta.
Elko Sta.
Carlin Sta.
Camp Halleck
Hastings Road
Franklin Valley
Ft Ruby
Hastings Pass
Piñon Range
Reese R.
Kobeh Val.
WHITE PINE DISTRICT
Hamilton
Union Peak
Thousand Sp. Valley
Pilot Kn.
Preuss L.
Pahranagat
Alkali
Hico
Logan
Crystal Sp.
Elys L.
Beaverdam Cr.
Muddy R.
St Joseph
St Thomas
Callville
Hardyville
Mohave City
Ft Mohave
Olive City
NEVADA
CALIFORNIA
PACIFIC OCEAN

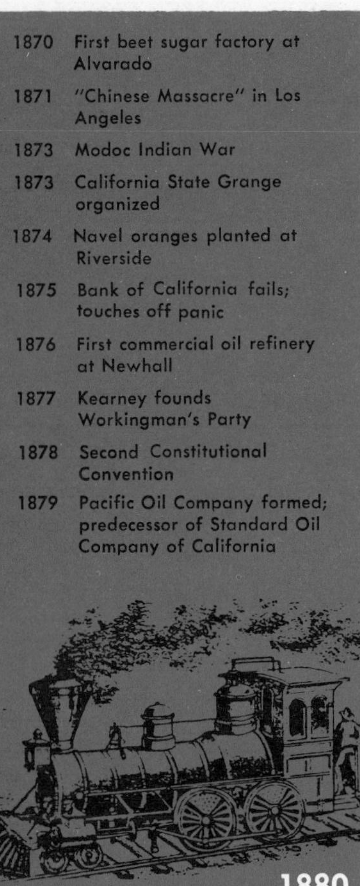

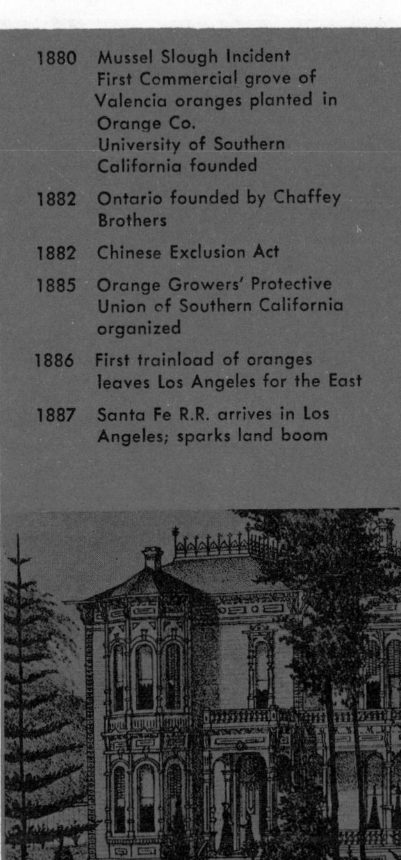

RAILROADS AND SETTLED AREAS

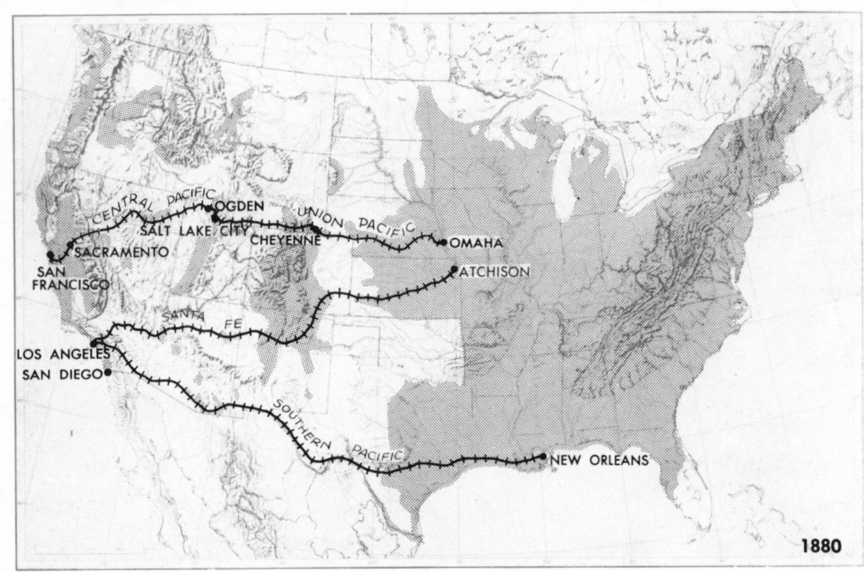

San Francisco Bay region and built southward through the Central Valley to southern California. They began operating as the Southern Pacific Railroad Company after 1865. In the process of expansion they absorbed the smaller lines in California and acquired their land grants. In all, the Southern Pacific received about 11,588,000 acres of land in California. In 1875 the railroad reached southern California, and by 1881 it connected with the Santa Fe at Deming, New Mexico Territory. The line reached New Orleans in 1883.

The Atchison, Topeka and Santa Fe Railroad continued to build westward in the 1880's. It finally reached Los Angeles by way of San Bernardino in 1887 and ran a new track across the Tehachapis from Barstow to provide service to the Central Valley and the San Francisco Bay area. This ended the monopoly of the Southern Pacific Railroad and ignited a land boom which eventually led to the present dominant position of southern California.

First International Air Meet

First junior college established in California

Initiative, Referendum and Recall Act approved
First transport of airmail

Alien Land Law prohibits Japanese ownership of agricultural land
Wheatland Riot
Owens River Aqueduct brings water to San Fernando Valley

Panama Canal opens

Mt. Lassen erupts

1920

1921 Oil discovery at Signal Hill
Transcontinental air mail

1922 Colorado River Compact signed

1927 "Spirit of St. Louis" built in San Diego
Lindbergh flies Atlantic

1928 St. Francis Dam disaster; 385 lives lost

1930

1931 Construction of Hoover Dam begins

1933 Death Valley National Monument created

1933 Stockton Ship Channel completed

1934 Townsend Plan proposed

1934 Work on All-American Canal begins

1935 Construction of Shasta Dam authorized

1936 Oakland Bay Bridge opened

1936 Hoover and Parker Dams completed

1936 Discovery of "Drake's Plate"

1937 Golden Gate Bridge opens

1939 Defeat of pension schemes, "Thirty Dollars Every Thursday," or "Ham and Eggs"

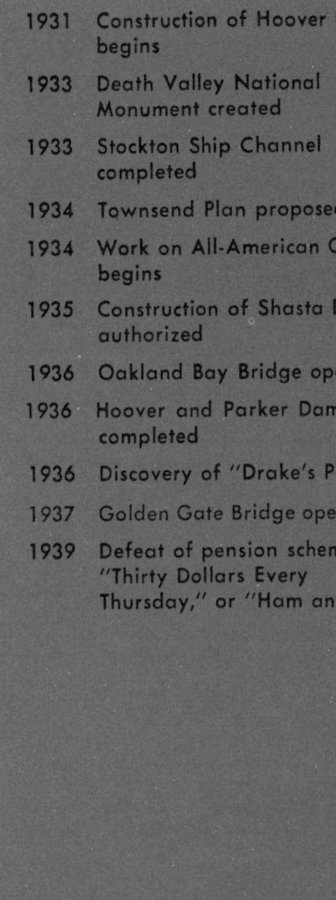

1940

1941 U.S. enters World War II

1941 Relocation of 112,000 Japanese from West Coast

1945 United Nations organized in San Francisco
End of World War II

1948 Mt. Palomar telescope put into use

1950

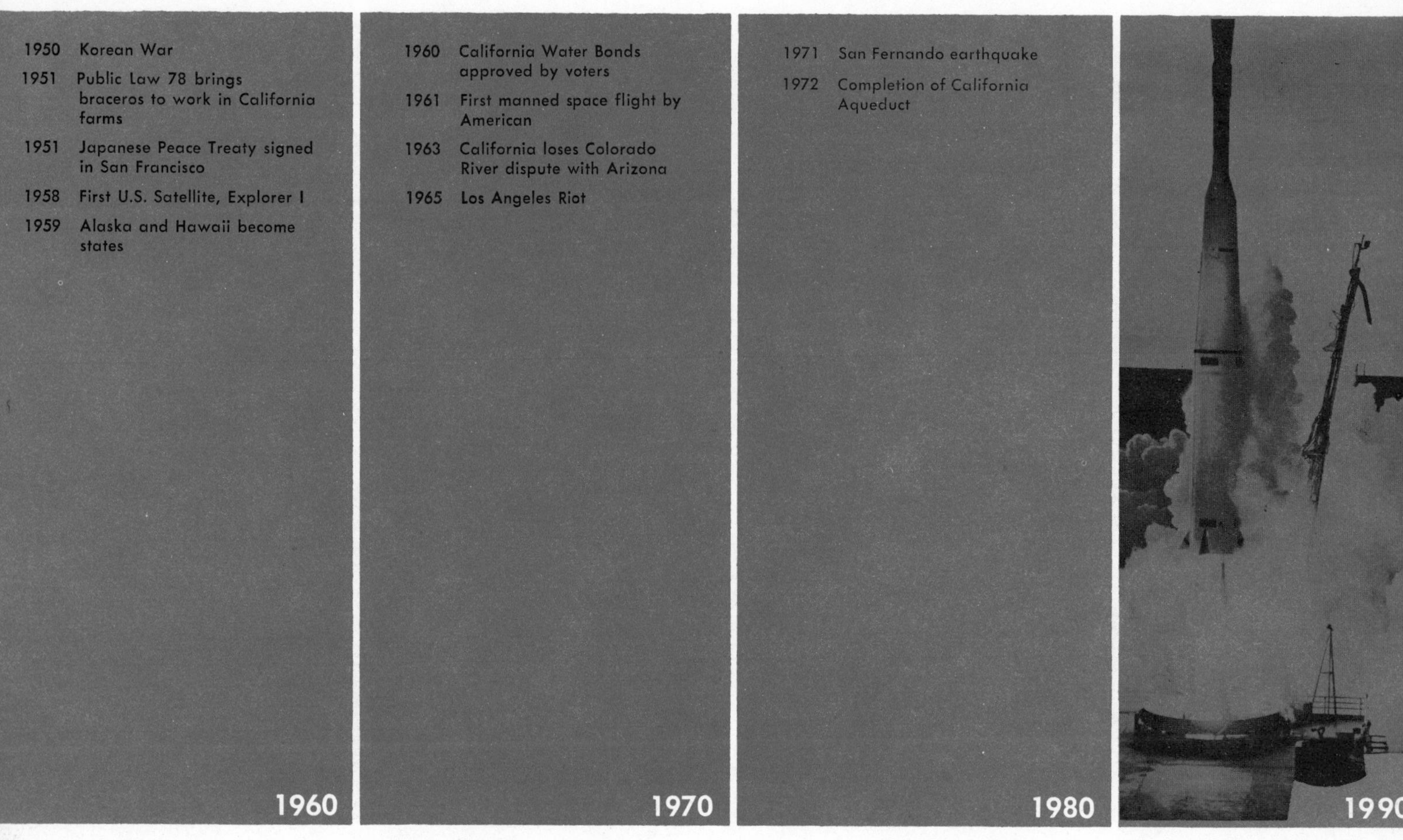

The "Boom of the Eighties" was sparked by rate wars between the two major railroads which drove passenger fares from the Midwest from $125 to as little as $1 and freight rates for California produce from $1,000 a car to $400 a car. People from all parts of the country poured into the state to partake of the scenery and the climate and to enjoy the benefits of the land boom. Dozens of new towns sprang up, and old ones mushroomed in size.

The lowered freight rates permitted a strong agricultural base to develop, and upon this base the whole California economy surged forward. The development of the automobile and truck, the aeroplane and the completion of the Panama Canal have all contributed to the binding of California more firmly to the rest of the nation. The development of manufacturing industries on a large scale in the period since World War II has contributed to the expansion of the California economy and has helped bring additional people to the state. Far-sighted governmental employees have assisted in the development of the water resources needed to keep the massive industrial and urban complexes supplied with their basic needs. Rapid population growth associated with the immigration of people from other parts of the United States and from other countries continues to characterize the California scene just as it did in the days of the "Forty-Niners."

Rapid population growth associated with the immigration of people from other parts of the United States and from other countries continued to characterize the California scene until recent years. With decreasing federal expenditures in the aerospace industries and consequent loss of jobs, and with further degradation of the national environment, the quality of life in the Golden State has become somewhat tarnished. Smaller numbers of people are migrating to California and larger numbers are fleeing the problems created by the tremendous post-war population growth. Perhaps in the seventies California will have a respite from the problems created by never-ending growth.

PART III POPULATION AND URBAN GROWTH

POPULATION GROWTH: 1850-1970

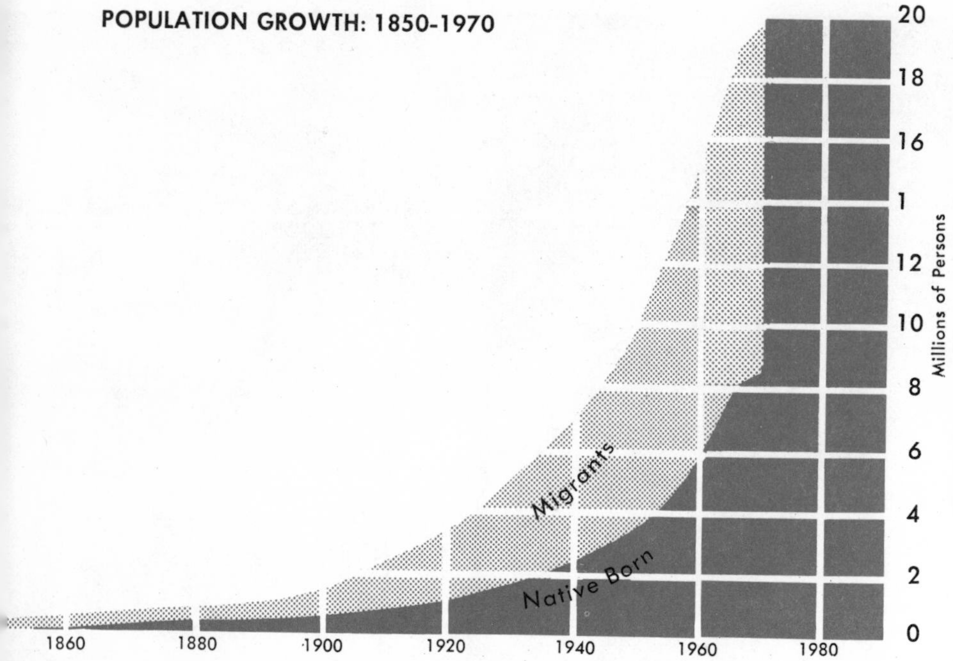

Migrants
Native Born
Millions of Persons
0 2 4 6 8 10 12 1 16 18 20
1860 1880 1900 1920 1940 1960 1980

ORIGIN OF THE POPULATION

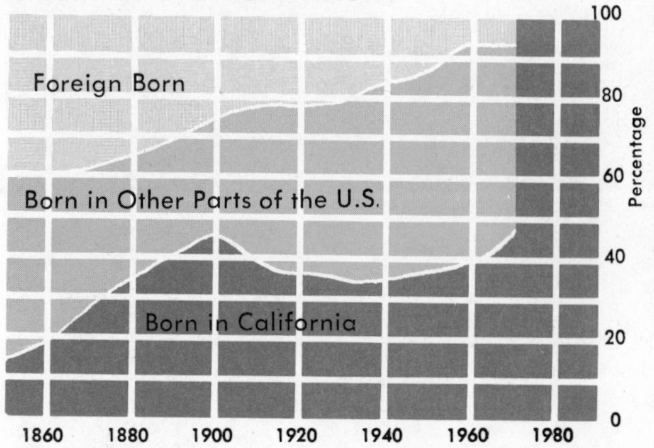

Foreign Born
Born in Other Parts of the U.S.
Born in California
Percentage
0 20 40 60 80 100
1860 1880 1900 1920 1940 1960 1980

ORIGIN — NATIVE BORN

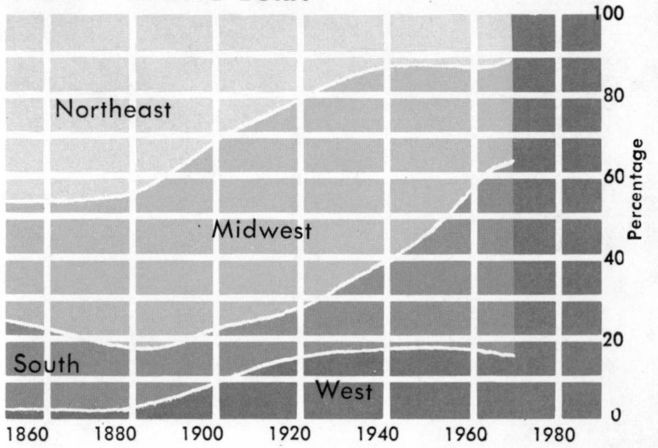

Northeast
Midwest
South
West
Percentage
0 20 40 60 80 100
1860 1880 1900 1920 1940 1960 1980

REGIONS OF THE UNITED STATES

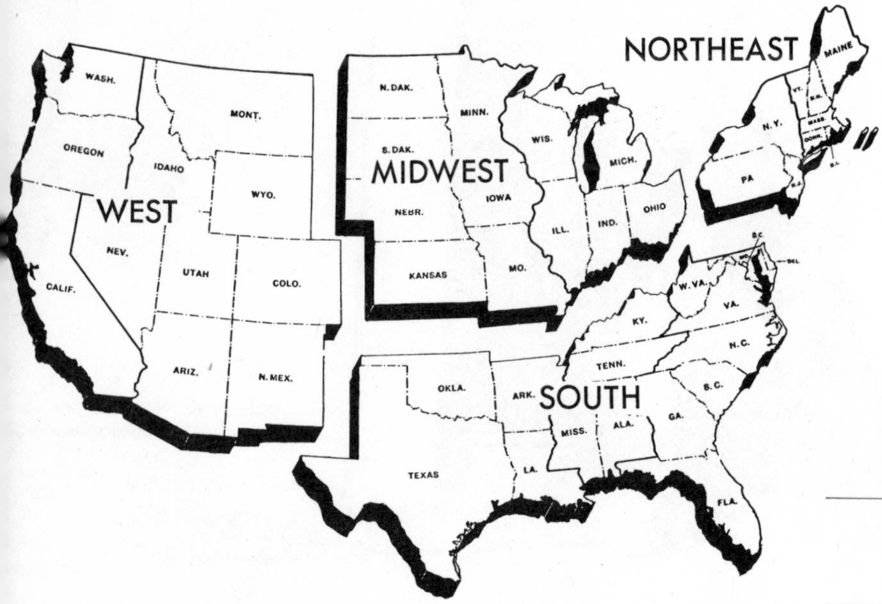

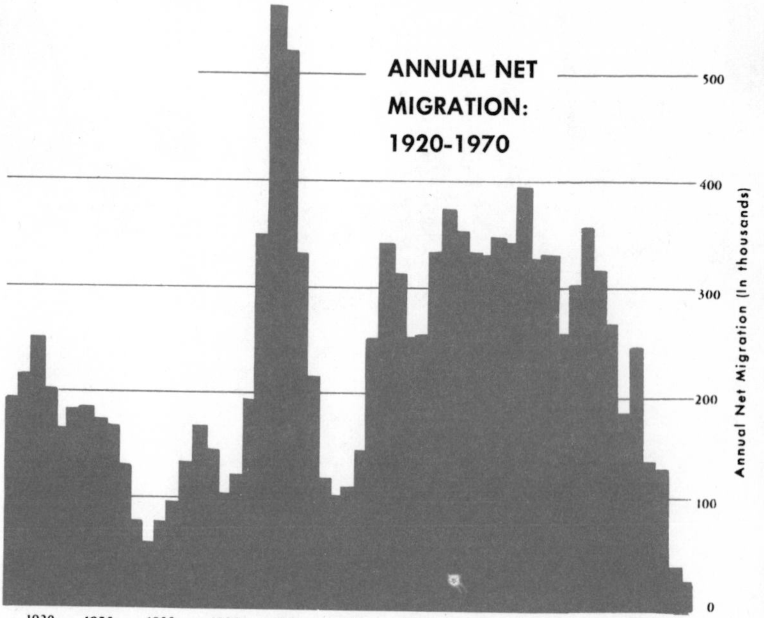

WEST
MIDWEST
NORTHEAST
SOUTH
WASH. OREGON IDAHO MONT. WYO. NEV. UTAH COLO. CALIF. ARIZ. N. MEX.
N. DAK. S. DAK. MINN. NEBR. IOWA KANSAS MO. WIS. MICH. ILL. IND. OHIO
MAINE N.Y. PA
OKLA. TEXAS ARK. LA. MISS. ALA. TENN. KY. GA. FLA. S.C. N.C. VA. W. VA. DEL.

ANNUAL NET MIGRATION: 1920-1970

Annual Net Migration (In thousands)
0 100 200 300 400 500
1920 1925 1930 1935 1940 1945 1950 1955 1960 1965 1970

POPULATION

California's population of almost twenty million people is larger than that of one hundred of the world's nations and larger than that of any of the other states in this nation. To support this population Californians have developed one of the most productive agricultural economies in the world, and its industrial and research firms have obtained a larger share of the federal space and military contracts than any other state. In spite of all these efforts, however, the people of the state face problems of unemployment and poverty which threaten to dwarf those of any other region. Most of these problems are associated with the rapid growth of population in recent years.

The problem of assimilating newcomers to the state is an old one. From the time of the discovery of gold, the growth of population in California has been characterized by the in-migration of large numbers of people from other

ORIGIN — FOREIGN BORN

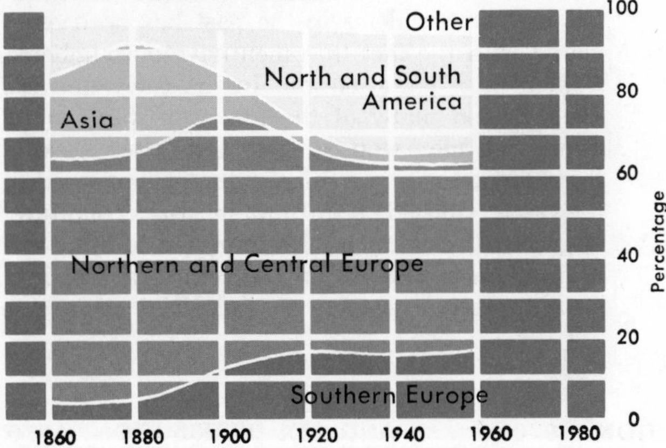

POPULATION DENSITY: 1850-1965

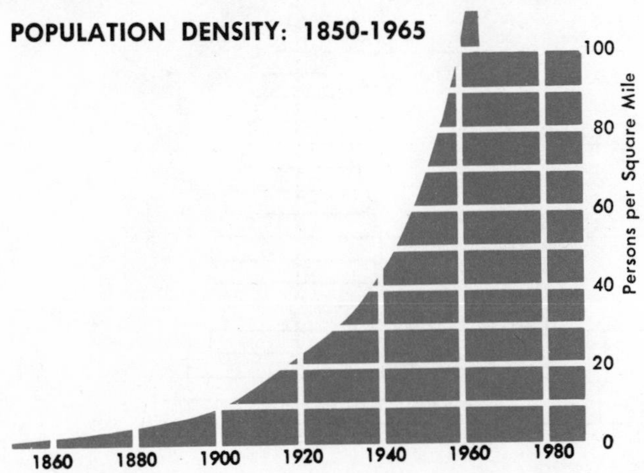

POPULATION GROWTH: MIGRATION AND NATURAL INCREASE

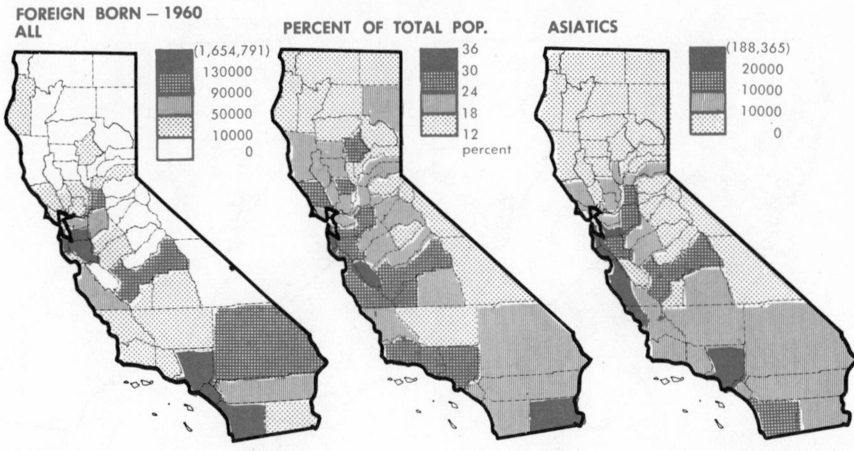

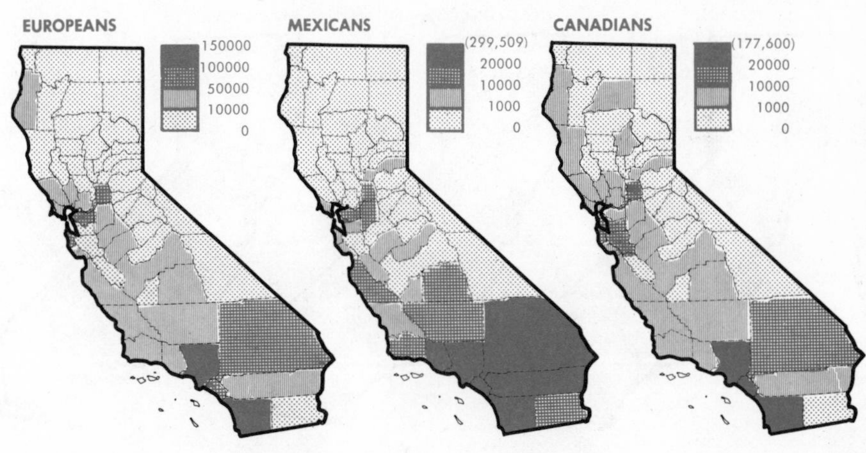

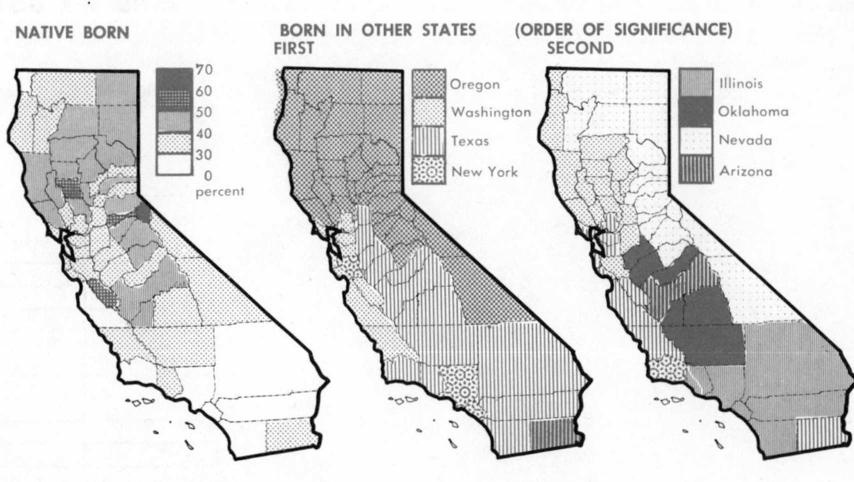

AGE OF POPULATION — 1970

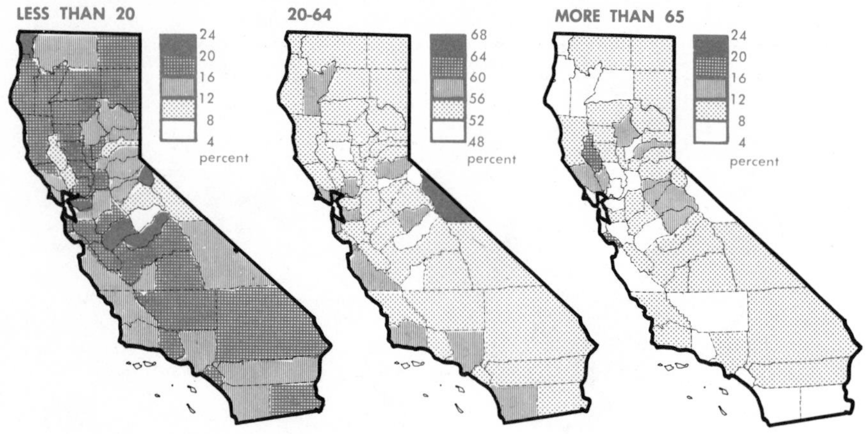

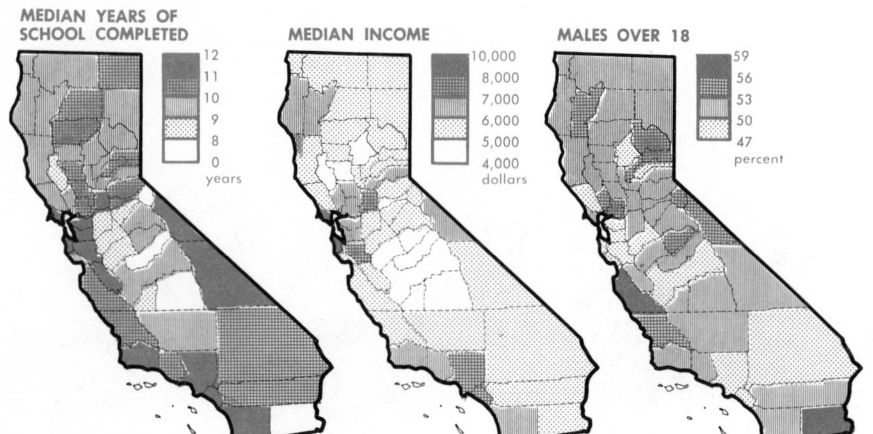

states and other countries. In no decade since becoming a state has the natural increase exceeded the growth due to migration.

People have come to California from all parts of the world, and the result has been a highly cosmopolitan population. There have always been a large number of foreign-born migrants. The bulk of the migrants have come from Europe although an unusually large number of them entered California from Mexico, China and Japan.

Immigrants from other states have added flavor and character to different parts of the state at different periods of time. During the Gold Rush many of the migrants came from the Northeast, and New England architecture and New England place names are to be found in the places where they settled. In time, more individuals from the Midwest and the South found their way to California and settled in the San Joaquin Valley and in the Los Angeles area where they have placed their distinctive imprint upon the land.

One other aspect of population growth which has created problems in the past has been associated with the fact that men have outnumbered women by a large margin. In the early decades of statehood many of the migrants were young unmarried men who came to California seeking their fortunes in the mines. As the economy of the state shifted from mining to agriculture, large numbers of men were imported to build the railroads and to work on the farms. It has only been in recent years with industrial and com-

CALIFORNIA AGE AND SEX DISTRIBUTION: 1970

U.S. AGE AND SEX DISTRIBUTION: 1970

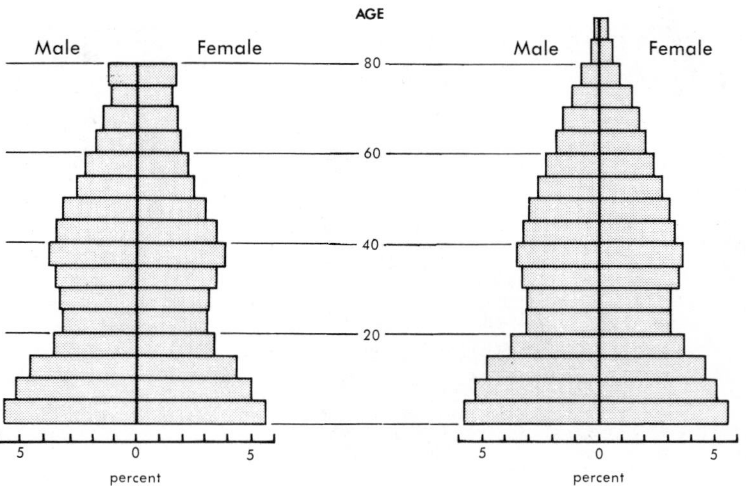

mercial development associated with urban growth that there have been as many women as men living in California.

Minority groups have always represented a large segment of the population, and Californians have not always recognized their rights and legitimate aspirations nor has there been agreement on the sort of public action necessary to meet problems arising from conflicting interests. From the middle 1850's to the 1880's one out of every ten Californians was Chinese. Restrictive laws and mistreatment soon eliminated them as a significant segment of our society. Similar treatment has kept the Japanese from achieving any great numerical strength. The Indians, Italians, Irish and Filipinos represent significant minorities found in the state in the past. Today, large numbers of Mexicans and Negroes represent people who are having a difficult time achieving all the advantages associated with living in California. Society must assist them to achieve their goals of equal opportunity in education, employment and housing.

A surprisingly large proportion of the people of California have lived in her cities during the years of American occupation and settlement. It is true that the initial surge of population swept into the foothills of the Sierra Nevada where gold was discovered, but by the late 1850's people were crowding into San Francisco. That city continued to dominate the state until southern California began to attract large numbers of people. In 1870 the growth rate of Los Angeles exceeded that of San Francisco and shortly after

CALIFORNIA AGE AND SEX DISTRIBUTION: 1960 and 1980

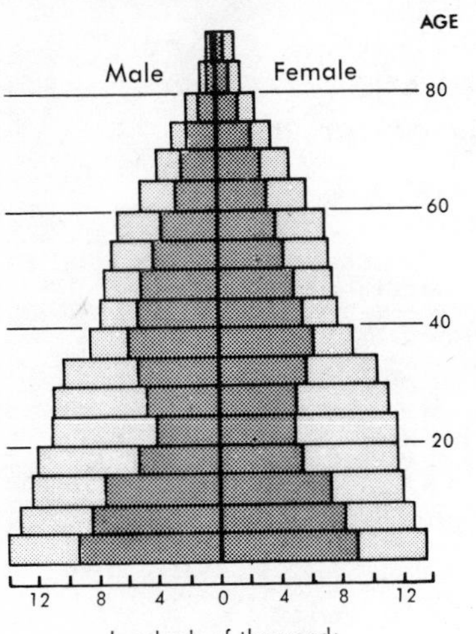

COMPOSITION OF THE POPULATION

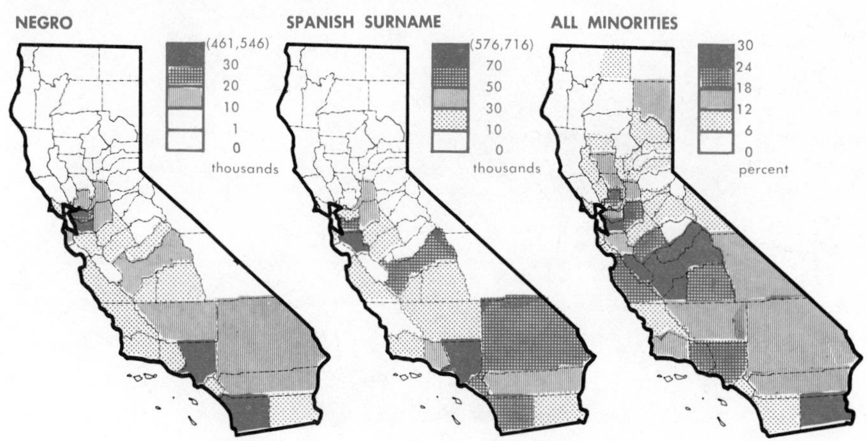

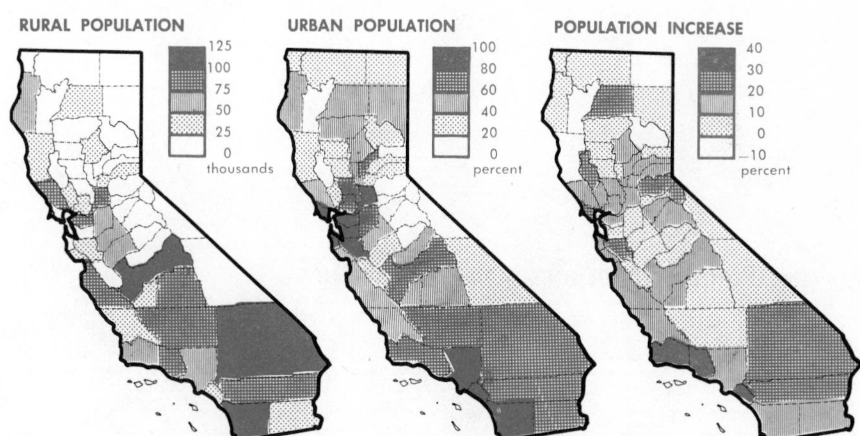

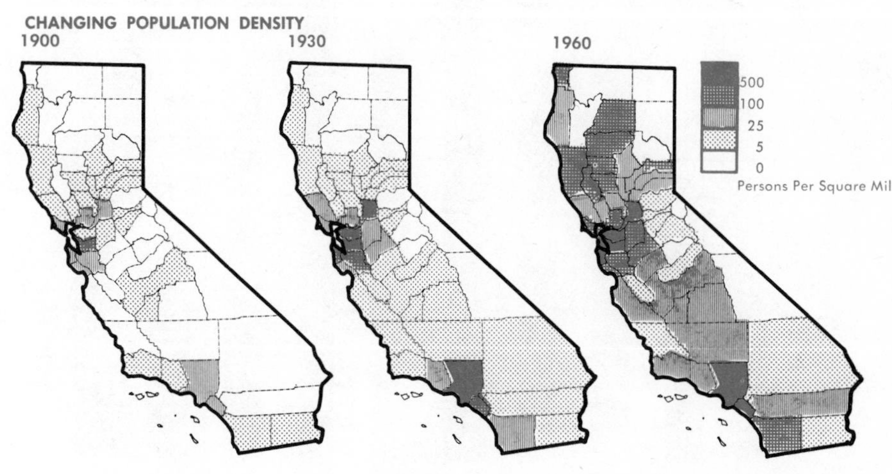

Los Angeles Civic Center.

1920 the population of Los Angeles became larger than that of San Francisco. In 1860 only a little over one-fifth of our population lived in towns and cities. Today, over 90 per cent of the population of California is to be found in her cities; three-fourths of them in the San Francisco, Los Angeles and San Diego metropolitan areas. Only New Jersey and Rhode Island are more highly urbanized. As our air has become polluted, our parks filled and our highways choked, Californians have become more aware of each other's presence and of the need to find solutions to the problems generated by metropolitan growth. Planning at the local, regional and state level is needed to provide for this and future generations.

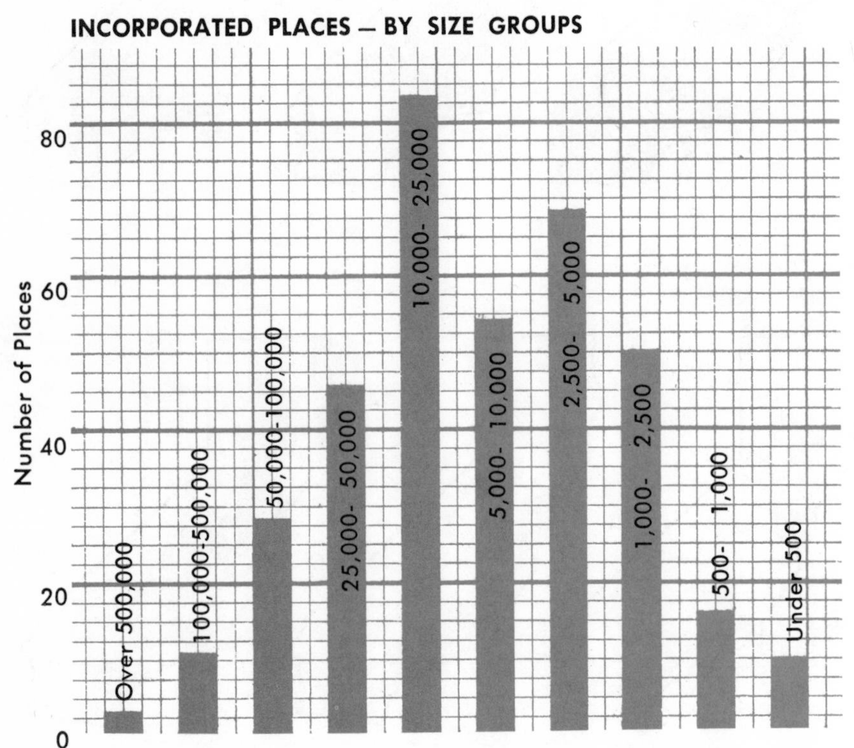

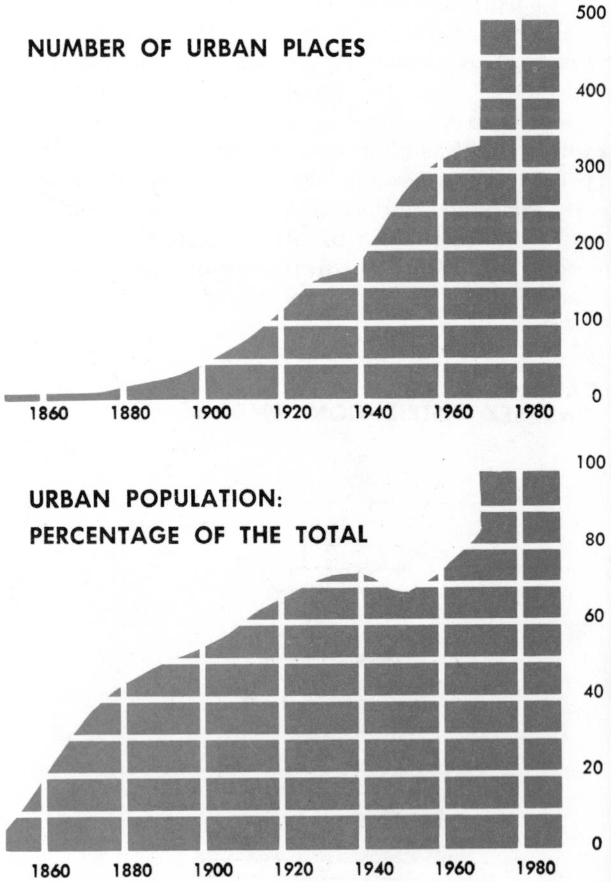

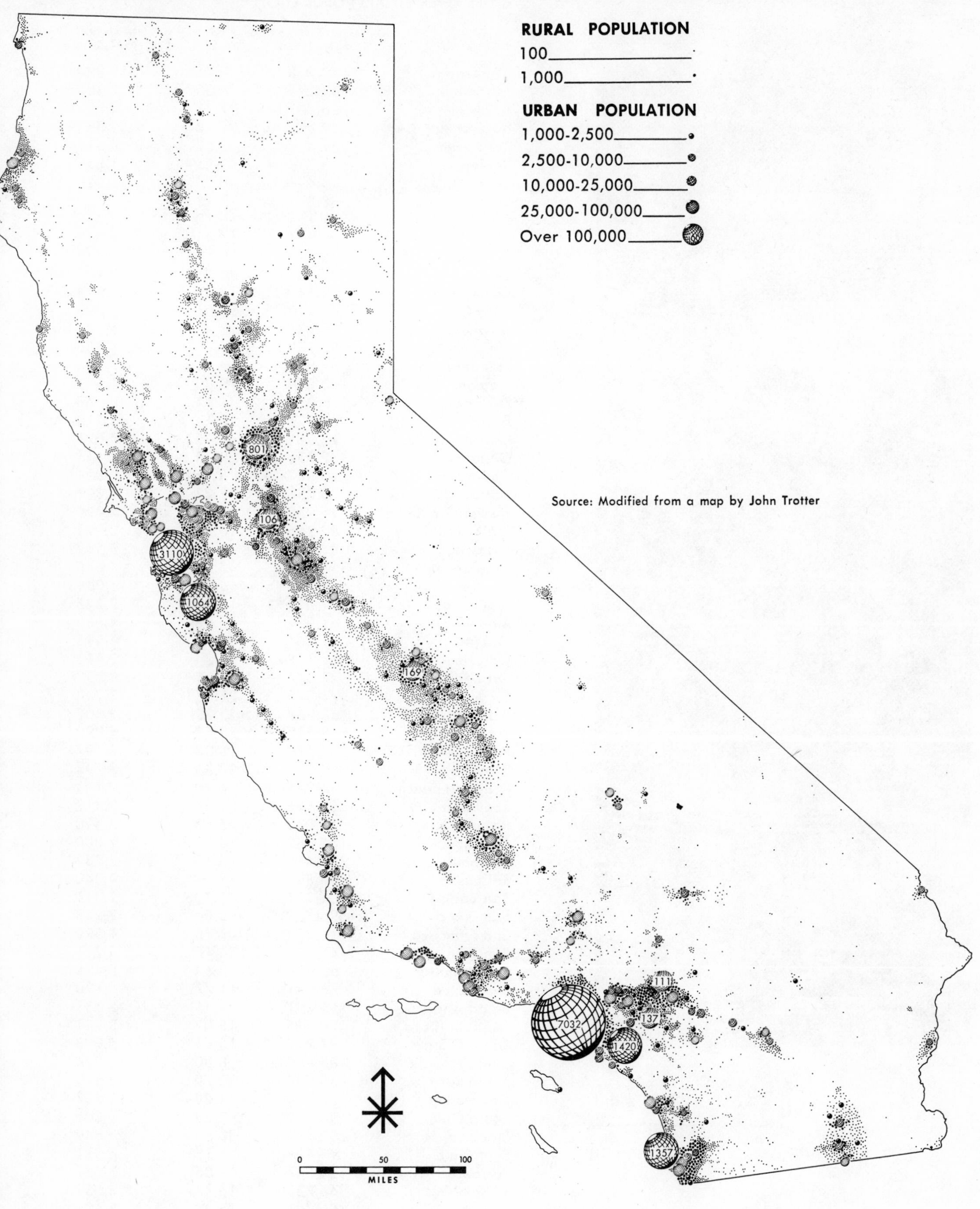

Source: Modified from a map by John Trotter

CITIES OVER 50,000 POPULATION*

CITY	POPULATION	AREA SQ. MI.	CONCENTRATION PER SQ. MI.
Alameda	70,968	7.75	10,270
Alhambra	62,125	8.60	7,207
Anaheim	166,701	34.80	4,732
Bakersfield	69,515	25.12	2,387
Bellflower	51,454	5.82	9,519
Berkeley	116,716	9.81	11,818
Buena Park	63,646	10.75	5,920
Burbank	88,871	17.13	5,130
Carson	71,150	16.10	4,419
Chula Vista	67,901	15.45	4,399
Compton	78,611	10.11	7,715
Concord	85,164	25.05	3,405
Costa Mesa	72,660	15.37	4,728
Daly City	66,922	10.01	6,690
Downey	88,445	13.43	6,585
El Cajon	52,273	11.99	4,356
El Monte	69,837	9.22	7,754
Fremont	100,869	99.00	1,019
Fresno	165,972	41.91	3,960
Fullerton	85,826	23.13	3,710
Garden Grove	122,524	17.45	7,021
Glendale	132,752	29.38	4,615
Hawthorne	53,304	5.21	10,391
Hayward	93,058	38.82	2,397
Huntington Beach	115,960	25.83	4,491
Inglewood	89,985	10.08	8,921
Lakewood	82,973	9.69	8,562
Long Beach	358,633	46.99	7,630
Los Angeles	2,816,016	466.36	6,038
Modesto	61,712	15.44	3,825
Mountain View	51,092	10.90	4,682
Oakland	361,561	54.82	6,595
Ontario	64,118	23.74	2,701
Orange	77,374	17.75	4,359
Oxnard	71,225	21.93	3,248
Palo Alto	55,966	23.30	2,402
Pasadena	113,327	22.67	4,999
Pico Rivera	54,170	8.33	6,497
Pomona	87,384	23.41	3,734
Redondo Beach	56,075	6.10	9,183
Redwood City	55,686	22.67	2,412
Richmond	79,043	25.80	3,490
Riverside	140,089	70.02	2,000
Sacramento	254,413	93.10	2,732
Salinas	58,896	13.57	4,340
San Bernardino	104,251	45.12	2,311
San Diego	696,769	310.04	2,248
San Francisco	715,674	44.75	1,599
San Jose	445,779	138.47	3,211
San Leandro	68,698	12.64	5,435
San Mateo	78,991	11.77	6,730
Santa Ana	156,601	27.55	5,681
Santa Barbara	70,215	18.02	3,901
Santa Clara	87,717	15.46	5,998
Santa Monica	88,289	8.00	11,036
Santa Rosa	50,006	20.79	2,399
Simi Valley	56,464	4.98	11,267
South Gate	56,909	7.29	7,806
Stockton	107,644	30.33	3,549
Sunnyvale	95,408	21.13	4,511
Torrance	134,584	19.06	7,061
Vallejo	66,733	16.44	4,059
Ventura	55,797	15.82	3,526
West Covina	68,034	14.52	4,686
Westminster	59,865	10.68	5,608
Whittier	72,863	12.25	6,459

*Based on 1970 census data.

PART IV UTILIZATION OF RESOURCES

LAND USE

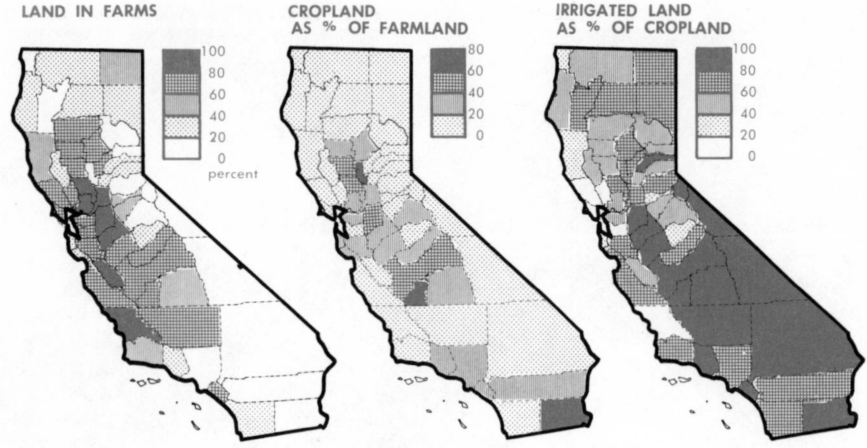

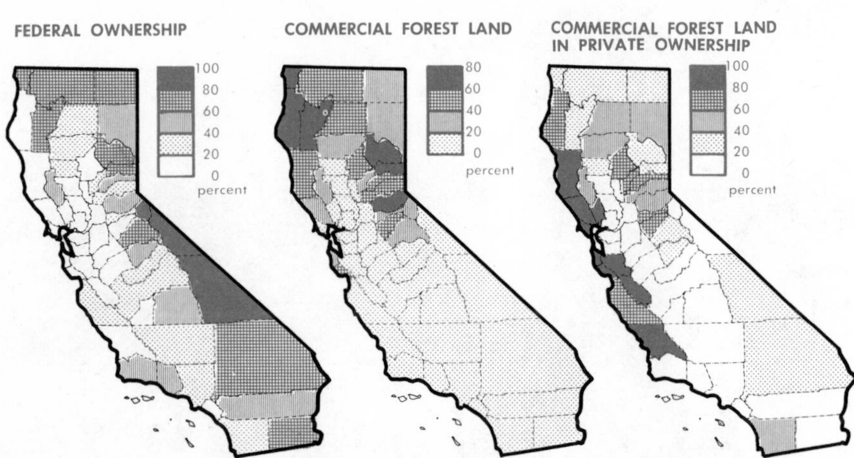

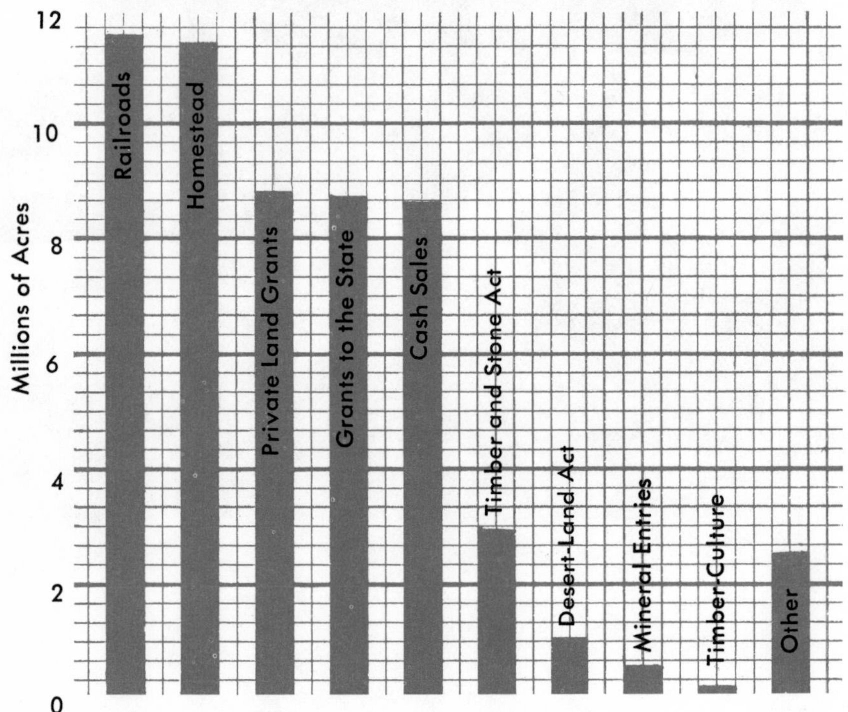

LAND USE

In the West there are great contrasts within the region and within the boundaries of each of the western states. Nowhere is this more noticeable than in California. Here rugged mountains stand in marked contrast to fertile flat-floored valleys; stark deserts present a different face than do humid coastal forests. Nowhere else has man done as much to modify the existing landscape.

Throughout California's history the changing patterns of land use have reflected the cultural background of the people living here, the limitations placed on their use of the natural environment, and the ever-increasing pressure of man on the land as the population has increased.

The Indians had a simple relationship with the land; they harvested its resources and accepted its limitations. During the Spanish and Mexican periods extensive land holdings, on which sheep and black Mexican cattle browsed, dominated the scene. Livestock ranching represented the major land use pattern. In the early years of the American period the banks and beds of the streams in all parts of California were searched for gold. As the mining boom subsided, agriculture replaced it as the dominant activity, and the demand for good agricultural land increased.

Today, the growth of the metropolitan areas has added a new dimension to man's use of the land, and the most rapid modification of the landscape is occurring in the urban areas. Within the city, houses are being replaced by apartments and office buildings, and factories consume an increasing amount of land. Blight strikes at the central core and along the rural fringe where land speculation disturbs the placid agricultural scene. Unchecked, it will destroy the beauty which has attracted countless people to the shores of the Pacific.

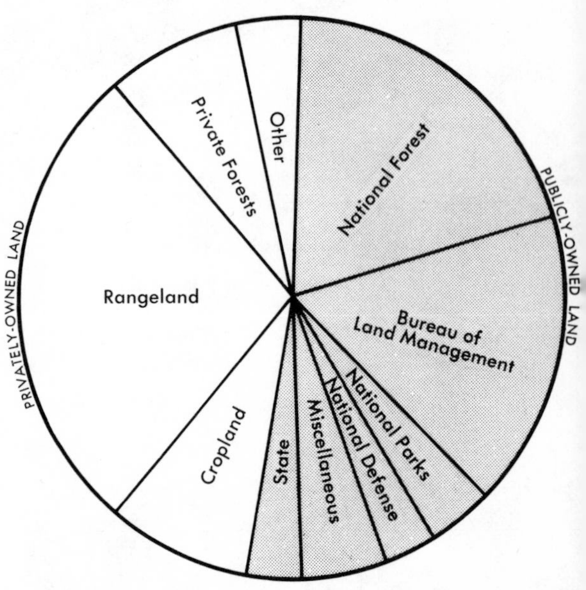

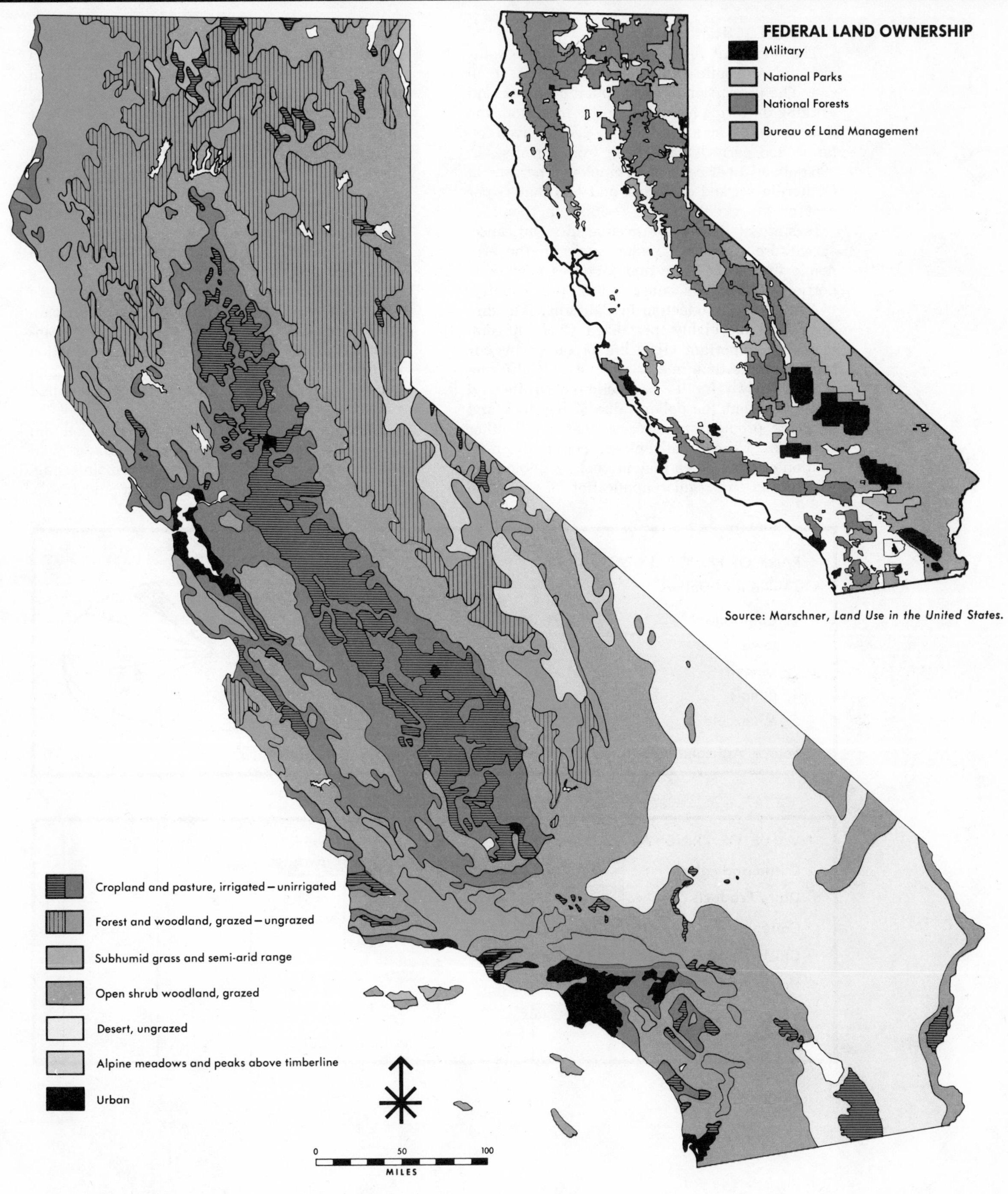

Source: Marschner, *Land Use in the United States.*

AGRICULTURE

The dominant role of agriculture in the political and economic affairs of California is at an end. The proportion of our population living and working on farms has steadily declined from the time when a majority of our people lived on farms and earned their living from farming. In 1965 about 5 per cent of all employed persons in California worked on farms, and less than 3 per cent of the state's people resided on them.

In spite of its relative decline in significance, agriculture remains a major factor in the economic life of the state and produces a sizeable portion of the food consumed in this country.

Agricultural production in California is highly diversified and highly specialized. Over 140 commercially important crops are produced by our farmers; more than in any other state. California is best known for its production of fruits and vegetables, but the dollar value of livestock and livestock products far exceeds that of all other categories. Climatic advantages permit the maturation of crops when they are not available elsewhere, and the regular application of irrigation water and fertilizer assures the farmer of higher yields of crops grown here than is true for regions dependent upon rainfall. In addition, there is a smaller crop loss to insects and disease since these plant enemies do not thrive in the dry heat of the California summer.

Technological advances and expanding local markets have helped to change the character of production. More meat, truck crops, eggs and dairy products are needed by the growing population of the state.

In the period prior to the entry of the Spanish into the American Southwest, Indians living along the Colorado River were the only native Californians practicing agriculture. In the Spanish and Mexican periods livestock ranching dominated the scene, but there was some production of dry-farmed wheat and barley and of fruits and vegetables on irrigated land at the pueblos and missions.

Diversification of agricultural production did not occur immediately with the influx of gold miners. The raising of livestock remained significant until the great drouths of the mid-1860's.

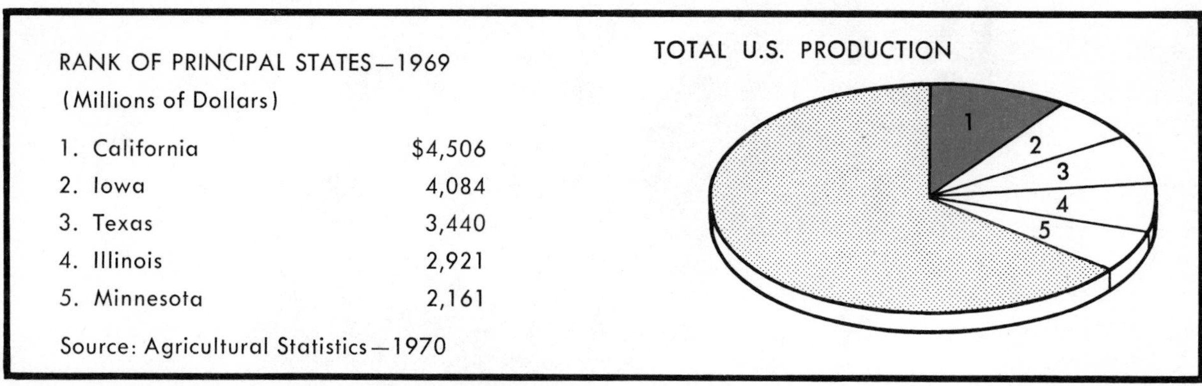

RANK OF PRINCIPAL STATES—1969
(Millions of Dollars)

1. California	$4,506
2. Iowa	4,084
3. Texas	3,440
4. Illinois	2,921
5. Minnesota	2,161

Source: Agricultural Statistics—1970

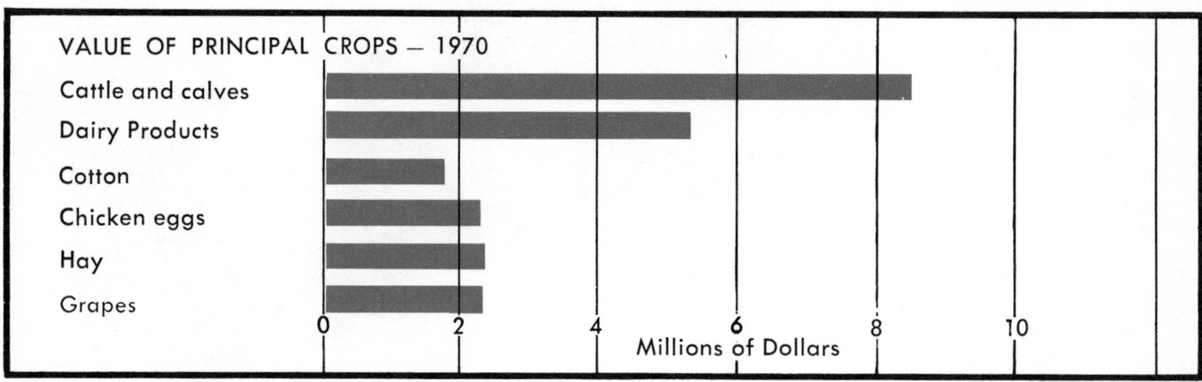

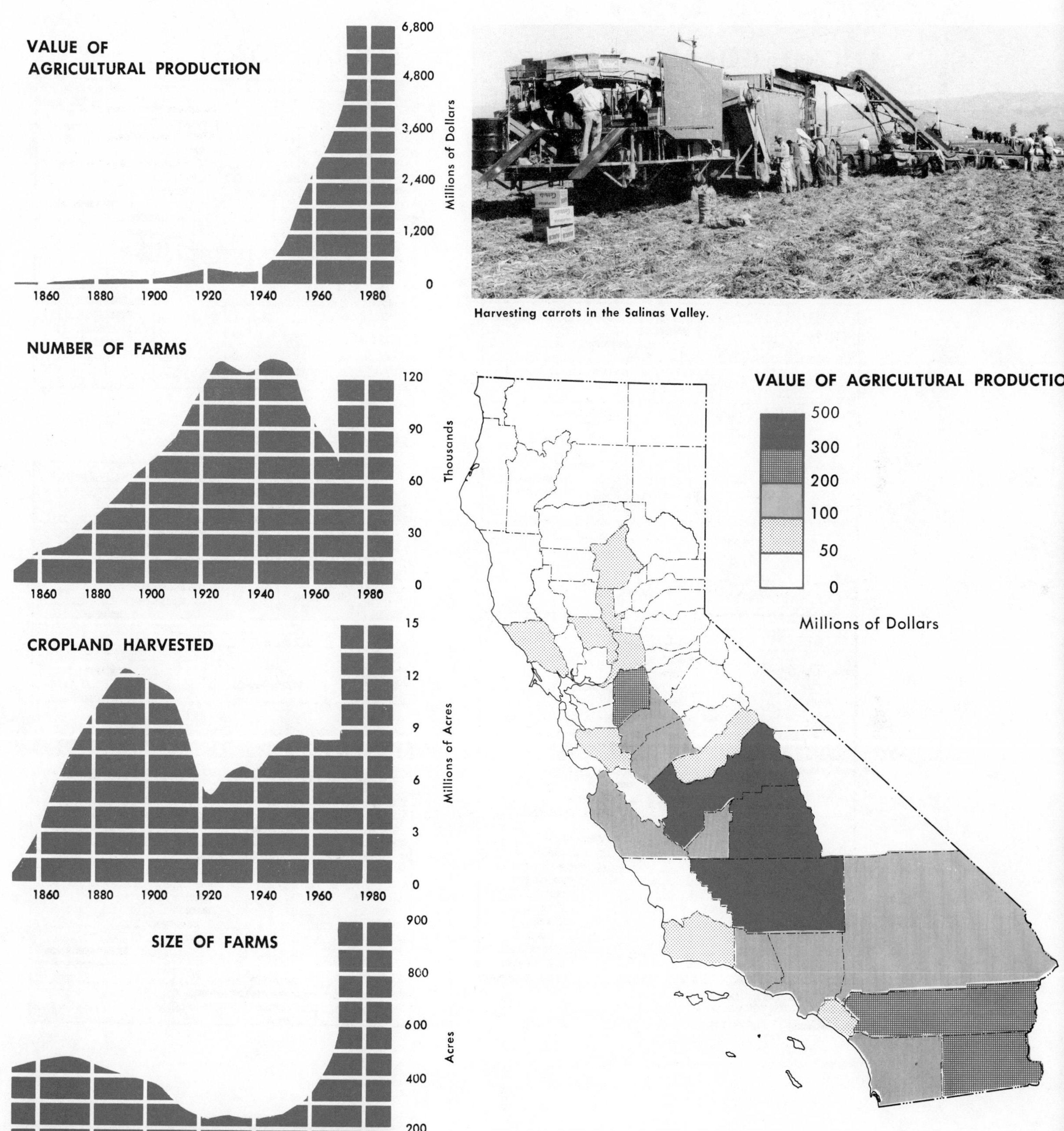

Harvesting carrots in the Salinas Valley.

CROP HARVEST PERIOD

NORTH COAST

CROP | JAN | FEB | MAR | APR | MAY | JUNE | JULY | AUG | SEPT | OCT | NOV | DEC

CHERRIES
BUSH BERRIES
SNAP BEANS
APPLES
PEARS
PRUNES
GRAPES, WINE
WALNUTS
ORCHARD PRUNING

SACRAMENTO VALLEY

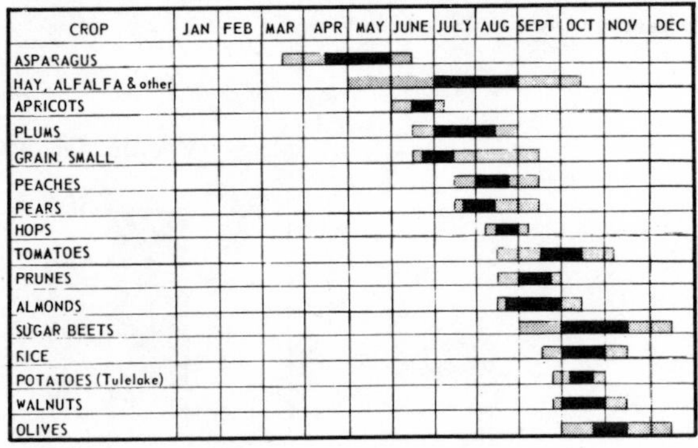

SAN JOAQUIN VALLEY

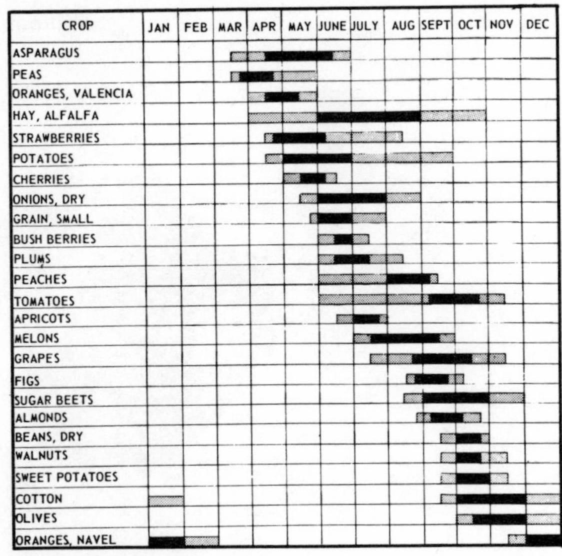

MAJOR WORK SEASONS ARE SHOWN BY BARS ACROSS THE CALENDAR

peak

HARVEST WORK

CENTRAL COAST

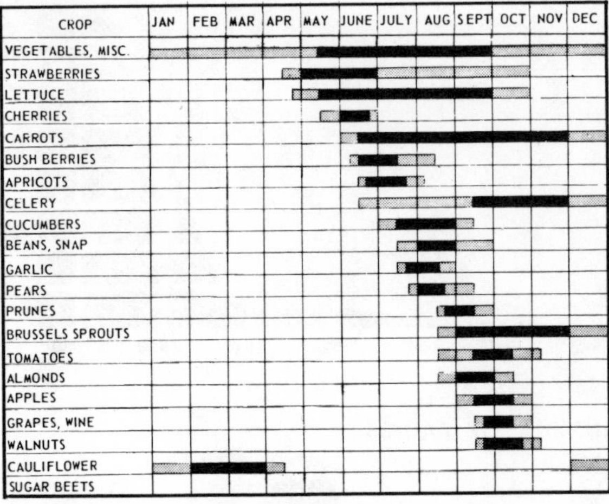

SOUTH COAST

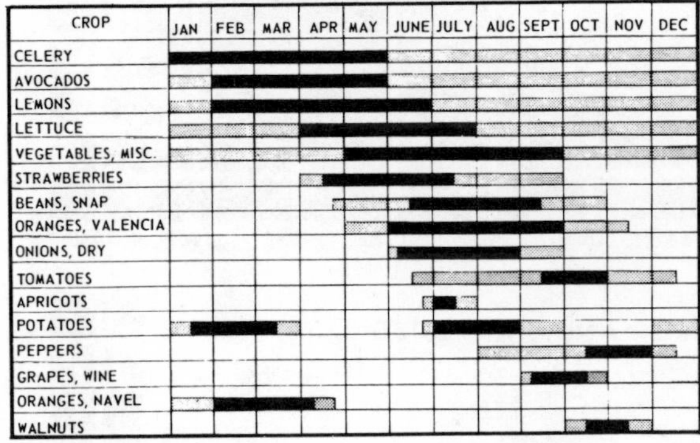

DESERT

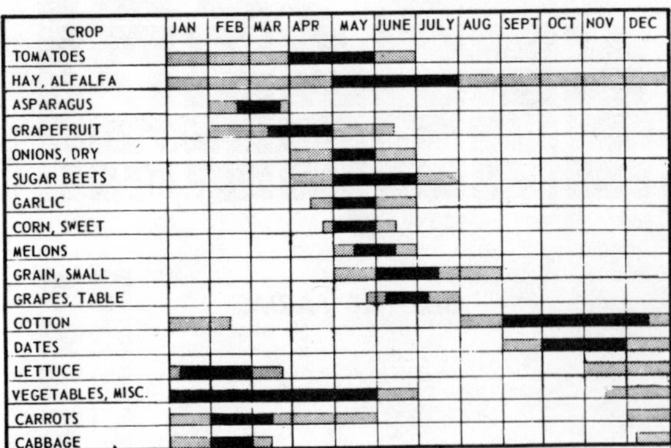

AREA — CROP LIST

MAJOR AREAS ■

OTHER AREAS ▒

Crop	Desert	South Coast	Central Coast	North Coast	San Joaquin Valley	Sacramento Valley
ALFALFA	▒				■	■
ALMONDS			▒		■	■
APPLES			■	■		
APRICOTS		▒	■		■	▒
ASPARAGUS	▒				■	▒
AVOCADOS		■				
BEANS, DRY					■	
BEANS, SNAP		■	■	▒		
BRUSSELS SPROUTS			■			
BUSHBERRIES			▒	▒	■	
CABBAGE	■					
CARROTS	■		■			
CAULIFLOWER			■			
CELERY		■	■			
CHERRIES			■	▒	■	
CORN, SWEET	▒					
COTTON	▒				■	
CUCUMBERS			■			
DATES	■					
FIGS					■	
GARLIC	▒		■			
GRAIN, SMALL	▒				■	■
GRAPEFRUIT	■					
GRAPES	▒	▒	▒	■	■	
HAY (SEE ALFALFA)						■
HOPS						▒
LEMONS		■				
LETTUCE	■	▒	■			
MELONS	■				■	
OLIVES					■	■
ONIONS	■	▒			■	
ORANGES, NAVEL		■			■	
ORANGES, VALENCIA		■			▒	
PEACHES					■	■
PEARS			■	■		■
PEAS					■	
PEPPERS		■				
PLUMS					■	▒
POTATOES, WHITE		▒			■	▒
POTATOES, SWEET					▒	
PRUNES			■	▒		■
RICE						■
STRAWBERRIES		■	■		▒	
SUGAR BEETS	■		▒		▒	■
TOMATOES	▒	▒	■		■	■
VEGETABLES, MISC.	▒	■	■			
WALNUTS		▒	■	▒	■	■

AREAS OF AGRICULTURAL PRODUCTION

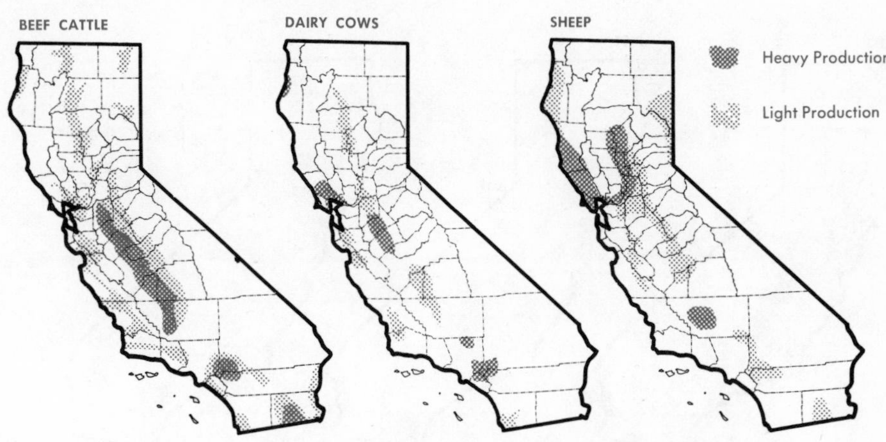

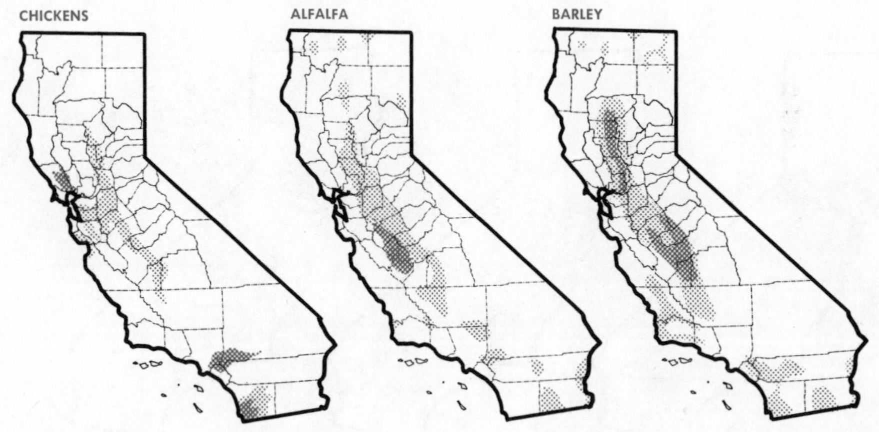

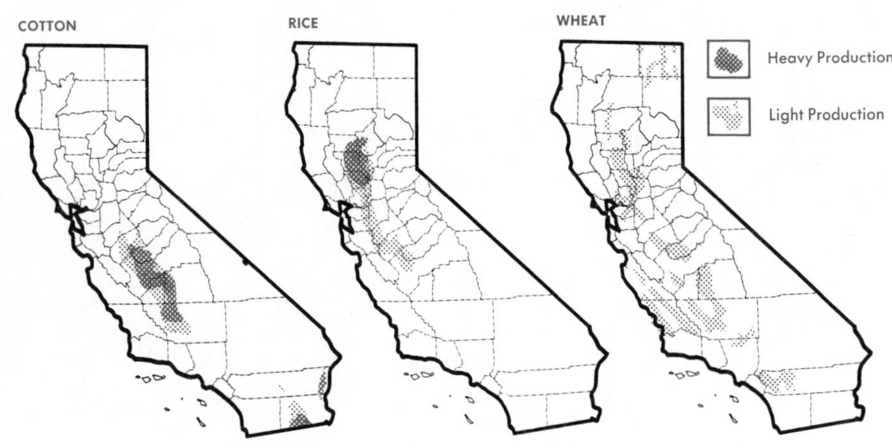

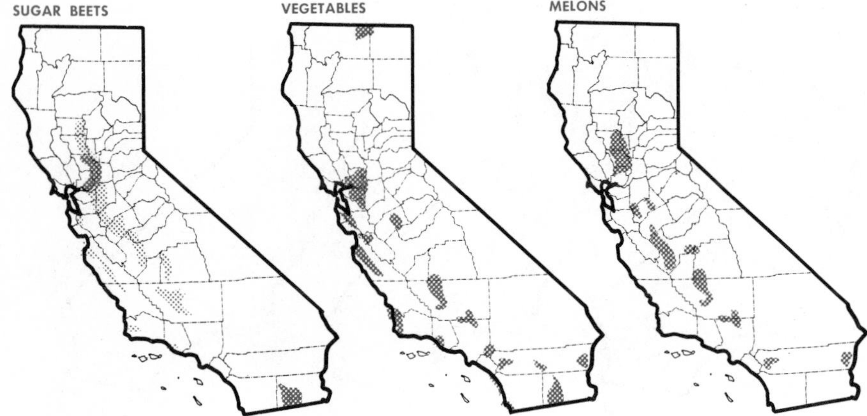

At that time the natural grasslands of the Central Valley were turned by giant plows, and bonanza wheat farming ruled supreme until the late 1880's when increased population coupled with increased availability of water, lower freight rates and falling prices on the world wheat market brought about the subdivision of the large wheat farms. From about 1890 to the present California agriculture has required large investments of manpower and capital as intensive use of the land has led to the specialization of production in various parts of California.

Requirements of increasingly large investments of capital and of greater technical skill led to the consolidation of land holdings in the agricultural areas of the state. The small farm has declined to vestigal significance only as the profit margin for the individual farmer has dwindled to almost nothing. In 1963 the average farm size in California was 400 acres, and only 3 per cent of the 99,000 farms contained two-thirds of the total acreage in farm land. The day of large-scale commerical production of specialty crops by large businesses is upon us. The average farm in California is worth about $150,000 and the investment needed to plant and harvest a crop is equally high.

Harvesting cotton.

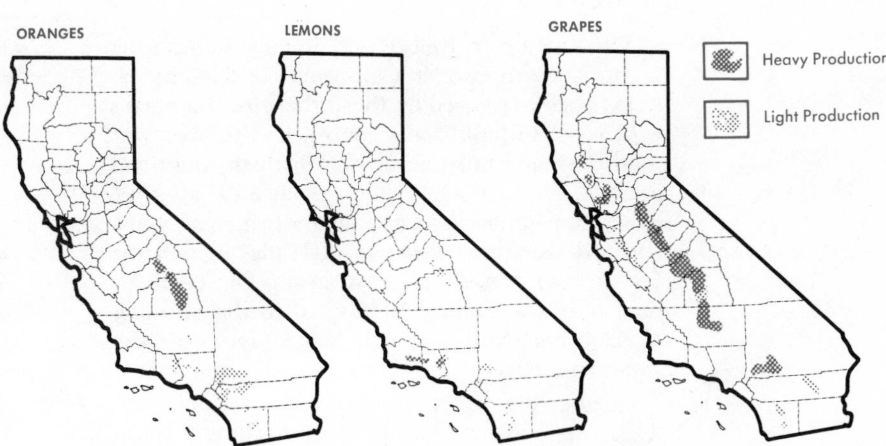

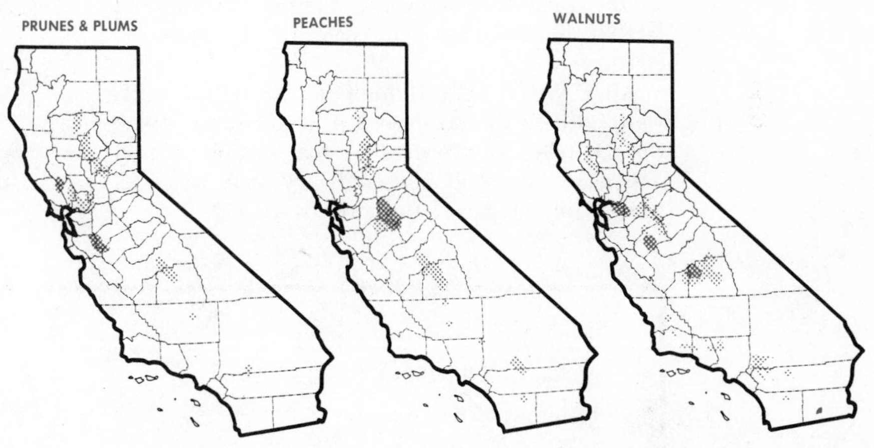

Rice fields in the Sacramento

FORESTRY

The cutting of timber and the manufacturing of lumber are two of California's oldest industries. Logs were used by the Indians of the northwest coast to build their houses, and lumber was used for many purposes in the Spanish-American period. The Russians had a sawmill at Fort Ross, and lumber was one of the principal items shipped to the Hawaiian Islands by Oliver Larkin, the first American Consul. The need for lumber in the Central Valley led to the building of Sutter's Mill on the south fork of the American River.

During the early years of the lumber industry the redwood forests to the north and south of San Francisco Bay were cut to provide wood to rebuild San Francisco each time that it burned. Gradually, lumber mills were established in the pine forests of the Sierra Nevada and farther north along the coast in the redwood country.

After World War II lumber and veneer mills were built in the northwestern counties to utilize the Douglas fir resource of that region. Rapid expansion of the lumber industry took place as the virgin forests of this area were cut.

California's production of wood and wood products ranks second to that of Oregon. The forest industries represent the most important segment of the economy of northern California and make a significant contribution to the economy of the state as a whole. Over one hundred thousand people are employed in the forest industries of the state.

Forests cover 42,500,000 acres of California land. Of this total 17,300,000 acres are classed as commercial forest land capable of producing saw timber and other forest products. The remaining 25,200,000 acres are reserved for parks and wilderness areas or are occupied by brush or woodland which is valuable as watershed and useful for grazing and recreation purposes.

As can be seen on the chart on the opposite page, our forest resource is being cut at a rate which is greater than is its rate of growth. In addition, present cutting practices are leading to rapid erosion of soil on the slopes of the hills and mountains of northwestern California. At the present rate of annual timber loss our forest resource will soon be exhausted, and future generations will be dependent upon synthetic products and lumber from plantations in the southern States.

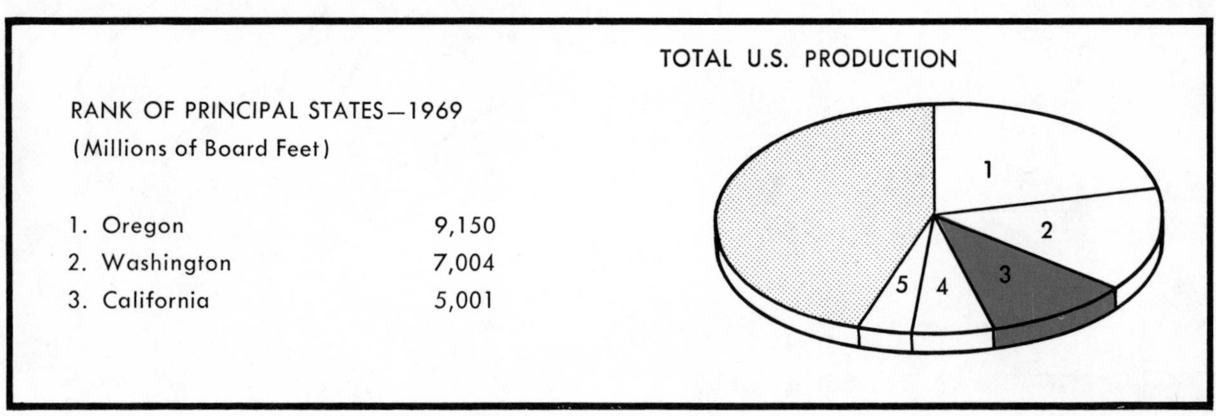

TIMBER VOLUME: 1953 AND 1963

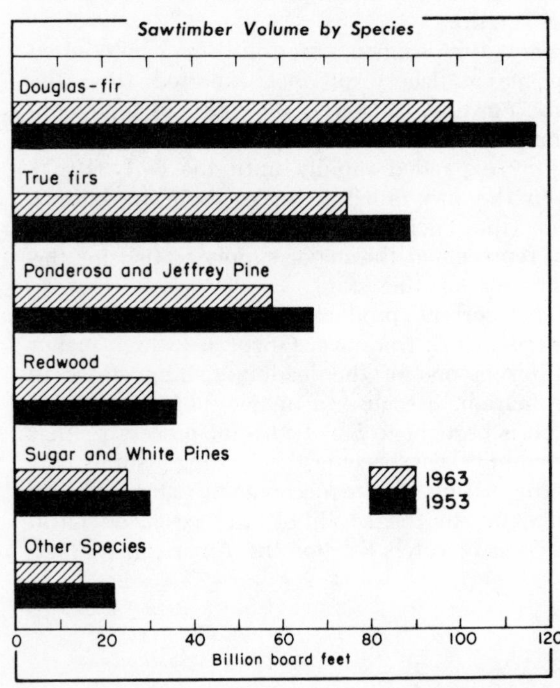

TIMBER PRODUCTION: 1850-1970

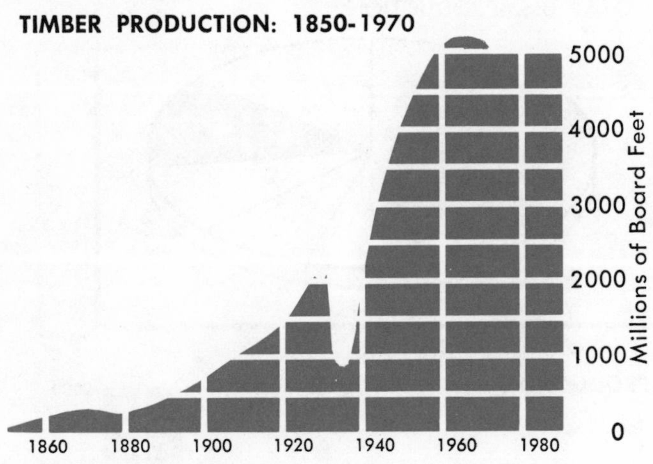

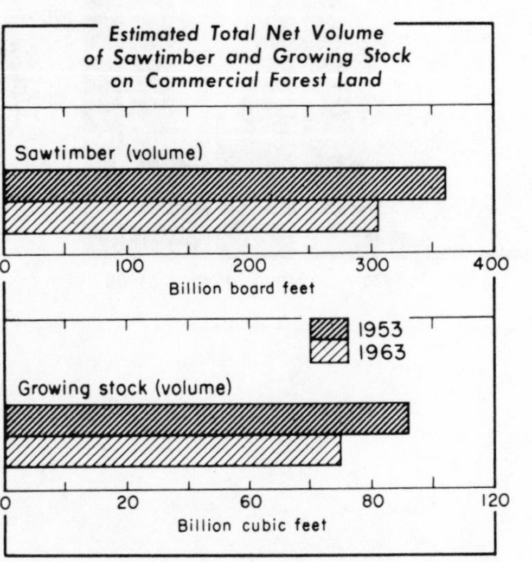

A typical lumber mill.

FOREST PRODUCTS

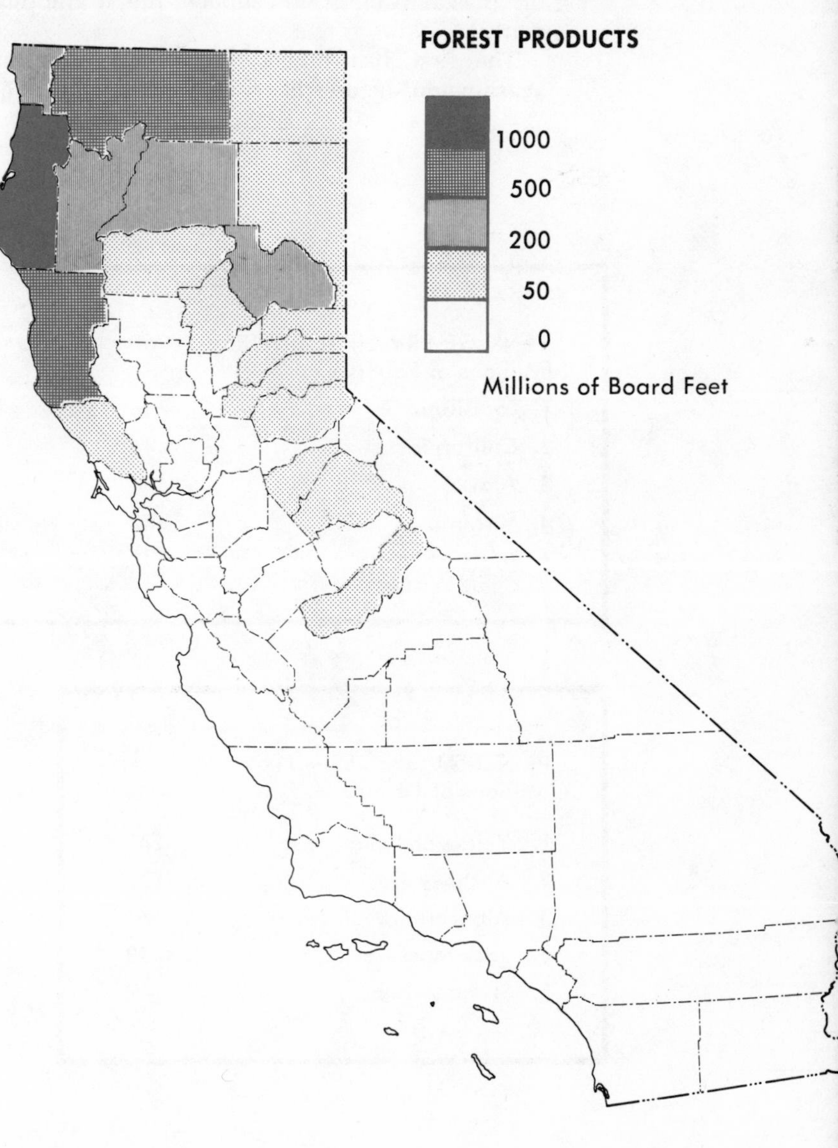

FISHERIES

From the time of the first settlement of California by the Indians the marine fisheries off her shores have been of considerable significance. Fish and mollusks provided a major supply of food for those people living along the coastal portions of the state, and dried and smoked fish sustained life during periods when other sources of food were scarce. Clam, abalone, and snail shells were used as articles of trade and adornment.

There was little fishing in coastal waters during the Spanish and Mexican times. However, by 1849 a small colony of Italians had established a small fishing industry in the San Francisco Bay area. Soon other fishermen using primitive nets and other gear plied their trade in coastal waters to supply the California market.

The first fish cannery began operation at Washington, across the river from Sacramento, in 1864. By the turn of the century canneries had been established in San Francisco, Monterey and San Pedro.

For a time salmon, cured in brine, represented the major fishery product exported from the state. Sardine, mackerel, sea bass and tuna were canned and sold in small quantities. The sardine fishery expanded rapidly until the early 1940's when the fish failed to make their appearance along the Pacific Coast. Since that time tuna has represented the major supply of fish for the canneries of the state.

The current production is over 600 million pounds of fish from over 60 species which makes California one of the leading fishing states in the nation. Because a major portion of this catch is brought to San Pedro for processing that community has become the leading fishing port in the nation. Large ocean-going tuna clippers range far to the south off the coast of Latin America to catch fish for the American market.

RANK OF PRINCIPAL STATES — 1969
(Millions of Pounds)

State	Millions of Pounds
1. Louisiana	748
2. California	446
3. Alaska	434
4. Virginia	382
5. Massachusetts	338

TOTAL U.S. PRODUCTION

1
2
3
4
5

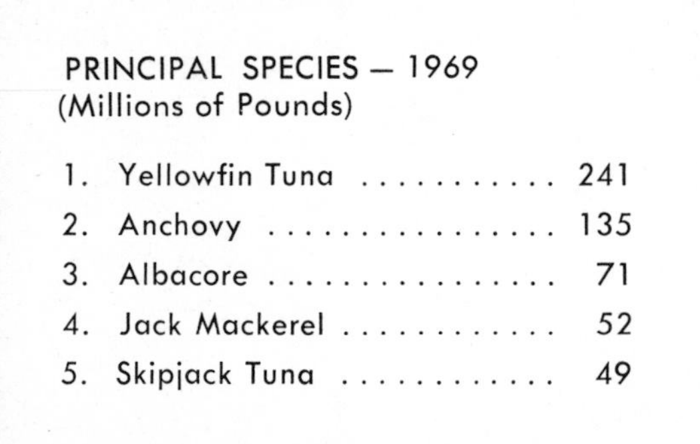
PRINCIPAL SPECIES — 1969
(Millions of Pounds)

Species	Millions of Pounds
1. Yellowfin Tuna	241
2. Anchovy	135
3. Albacore	71
4. Jack Mackerel	52
5. Skipjack Tuna	49

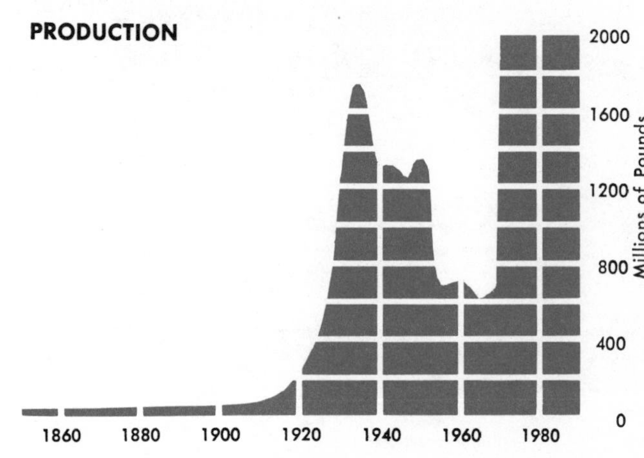

FISHING PORTS AND AREAS—1969

	Value	Thousand Pounds
EUREKA AREA	$ 7,298,147	45,270
SAN FRANCISCO AREA	2,378,956	8,901
MONTEREY AREA	1,654,516	24,501
SANTA BARBARA AREA	1,978,107	34,614
LOS ANGELES AREA	60,404,003	489,348
SAN DIEGO AREA	11,722,440	70,156

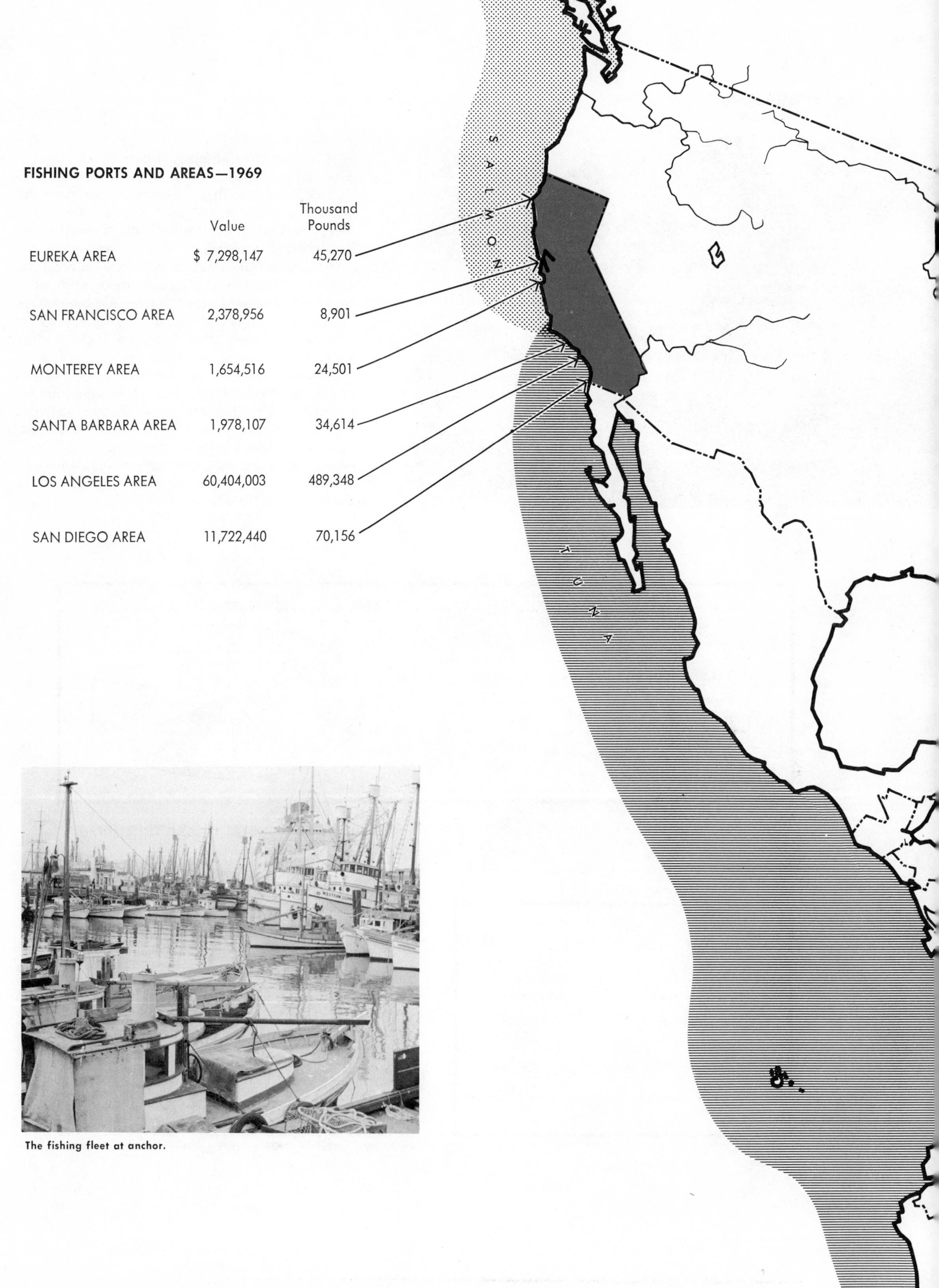

The fishing fleet at anchor.

Borax Mining

MINERALS

Gold brought California renown as a mineral-producing state and lured hundreds of thousands of people to seek their fortunes in this state. Although the mining of gold reached significant levels by 1852, it declined sharply thereafter and has virtually ceased.

Today, petroleum dominates the mineral industry in California although there are significant quantities of other minerals produced in the state. California has a greater variety of minerals than any other state and is the leading producer of many of these.

Minerals used in the construction industries are of particular significance. Adequate supplies of sand and gravel located on the perimeter of the major urban centers are essential to the builders of homes, office buildings, roads and bridges. Cement plants utilizing minerals located nearby are also to be found on the margins of the large metropolitan areas.

RANK OF PRINCIPAL STATES—1969
(Millions of Dollars)

1. Texas	$5,769
2. Louisiana	4,685
3. California	1,851
4. Oklahoma	1,091
5. Pennsylvania	976

TOTAL U.S. PRODUCTION

1 2 3 4 5

VALUE OF ALL MINERAL PRODUCTION: 1970

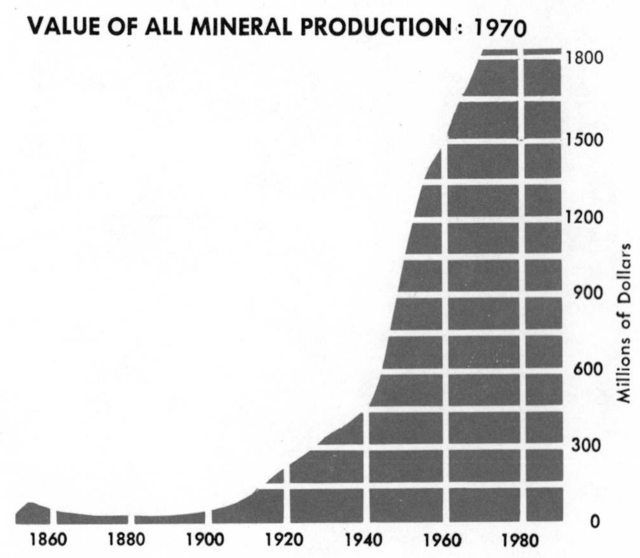

PRINCIPAL MINERALS—1970
(Thousands of Dollars)

1. Petroleum	$929,750
2. Natural Gas	194,436
3. Cement	174,376
4. Sand and Gravel	163,414
5. Boron	83,870

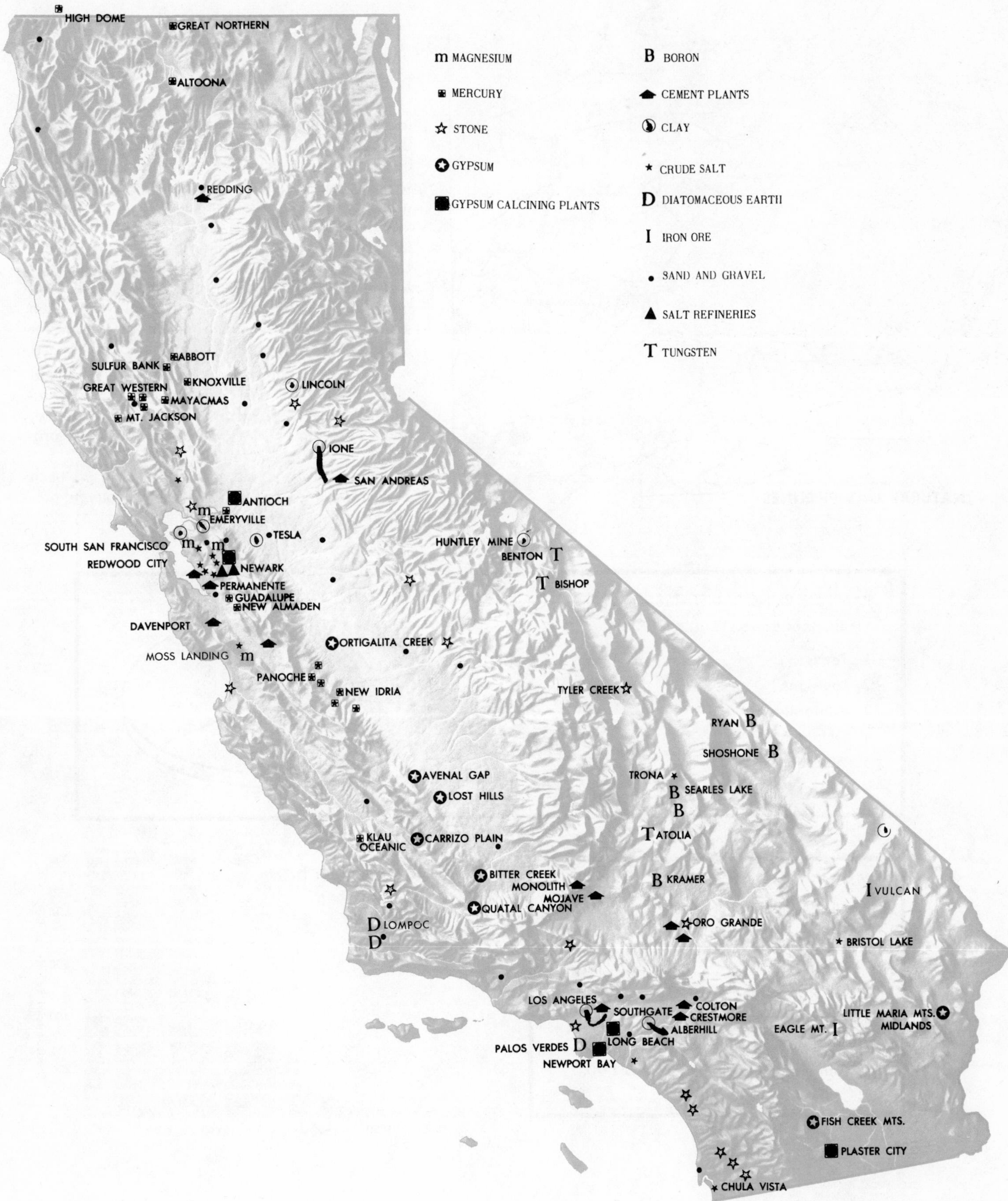
m MAGNESIUM
MERCURY
STONE
GYPSUM
GYPSUM CALCINING PLANTS
B BORON
CEMENT PLANTS
CLAY
CRUDE SALT
D DIATOMACEOUS EARTH
I IRON ORE
SAND AND GRAVEL
SALT REFINERIES
T TUNGSTEN
HIGH DOME
GREAT NORTHERN
ALTOONA
REDDING
SULFUR BANK
ABBOTT
KNOXVILLE
GREAT WESTERN
MAYACMAS
MT. JACKSON
LINCOLN
IONE
SAN ANDREAS
ANTIOCH
EMERYVILLE
TESLA
SOUTH SAN FRANCISCO
REDWOOD CITY
NEWARK
PERMANENTE
GUADALUPE
NEW ALMADEN
DAVENPORT
MOSS LANDING
ORTIGALITA CREEK
PANOCHE
NEW IDRIA
HUNTLEY MINE
BENTON
BISHOP
TYLER CREEK
RYAN
SHOSHONE
TRONA
SEARLES LAKE
ATOLIA
AVENAL GAP
LOST HILLS
KLAU
OCEANIC
CARRIZO PLAIN
BITTER CREEK
MONOLITH
MOJAVE
QUATAL CANYON
KRAMER
VULCAN
LOMPOC
ORO GRANDE
BRISTOL LAKE
LOS ANGELES
SOUTHGATE
COLTON
CRESTMORE
ALBERHILL
LITTLE MARIA MTS.
MIDLANDS
EAGLE MT.
PALOS VERDES
LONG BEACH
NEWPORT BAY
FISH CREEK MTS.
PLASTER CITY
CHULA VISTA

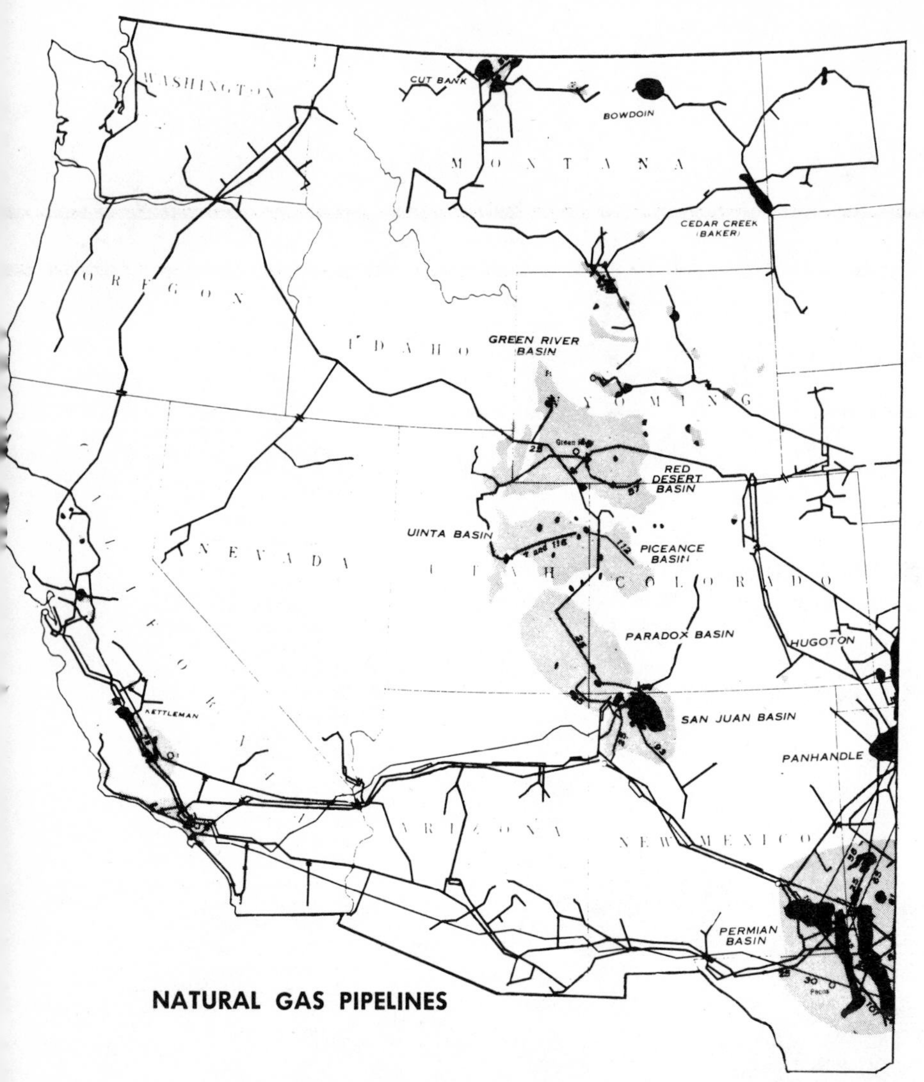

NATURAL GAS PIPELINES

PETROLEUM

For centuries the petroleum resources of California have been used by man. The Indians used the lighter oils as medicine while tar made their baskets water-tight and secured spearheads to wooden shafts. The Spaniards and Mexicans used it to patch their roofs.

When the Americans came to California, they ignored the oil oozing from the ground for at that time it had little value. It was not until technological advances created uses for the oil, and until methods of refining the crude petroleum were developed that men began to search for supplies of it.

The first commercial refinery was built at Newhall in 1877. The industry grew slowly until 1903 when there were 3,000 wells in the state, and California had become the number one producer of oil in the nation.

Since that time many new fields have been discovered and the value of the petroleum produced has increased many times.

RANK OF PRINCIPAL STATES—1969
(Millions of Barrels)

1. Texas	1,152
2. Louisiana	895
3. California	375
4. Oklahoma	225
5. Wyoming	155
U.S. Total Production	3,372

TOTAL U.S. PRODUCTION

1 2 3 4 5

PRINCIPAL FIELDS—1970
(Barrels per day)

1. Wilmington	224,135
2. Midway-Sunset	89,705
3. Kern River	69,369
4. Dos Cuadras	54,343
5. Huntington Beach	44,903

ANNUAL PRODUCTION: 1850-1970

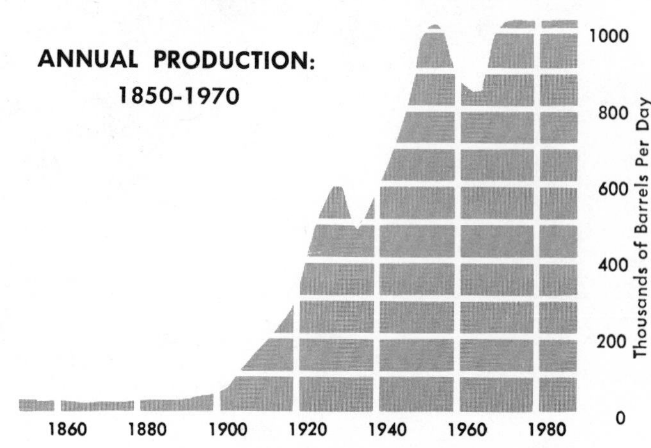

PETROLEUM

OIL FIELDS
1 La Honda Oil Creek
2 Moody Gulch
3 Flint Hills, Sargeant
4 Vallecitos, Ciervo
5 Chanley Ranch
6 Raisin City, San Joaquin
7 Helm
8 Riverdale
9 Coalinga
10 Guijarral Hills
11 San Ardo
12 Kettleman
13 Pyramid Hills
14 Blackwell's Corner, Alferitz Welcome Valley, Devils Den
15 Antelope Hills, McDonald
16 Lost Hills
17 Rio Bravo
18 Premier, Kern Front
19 Granite Canyon, Dorsey,
20 Round Mountain
21 Kern River, Kern Bluff, Ant Hill
22 Belridge
23 Cymrik
24 McKittrick, Belgian Anticline
25 Elk Hills
26 Coles Levee
27 Edison
28 Fruitvale
29 Canfield, Ten Sections
30 Paloma
31 Wheeler Ridge
32 Buena Vista Hills
33 Midway-Sunset
34 Tejon Hills, Tejon Grapevine
35 Russell Ranch, Cuyuma
36 Arroyo Grande
37 Santa Maria
38 Casmalia, Orcutt
39 Cat Canyon
40 Gato Ridge, Zaca Creek
41 Lompoc
42 Gaviota Conception
43 Goleta, Elwood
44 Santa Barbara
45 Summerland
46 Rincon
47 Sespe
48 Ramona
49 Ventura
50 South Mountain
51 Placerita
52 Aliso Canyon
53 Conejo
54 Montalvo
55 Oxnard
56 Salt Lake, Los Angeles
57 Beverly Hills
58 Inglewood
59 Montebello
60 Playa Del Rey
61 Potrero
62 Bandini
63 Whittier
64 El Segundo
65 Lawndale, Alondra
66 Rosecrans
67 Coyote, Santa Fe Springs
68 Olinda, Yorba Linda
69 Redondo
70 Richfield, Olive
71 Kraemer, Esperanza
72 Wilmington
73 Long Beach, Seal Beach
74 Huntington Beach, West Newport
75 Dos Cuadras
★ GAS FIELDS
1 Ord Bend
2 Marysville Buttes
3 Dunnigan Hills
4 Fairfield Knolls
5 Millar
6 Maine Prairie
7 Thornton
8 Rio Vista
9 Lodi
10 Kirby Hill
11 Suisun Bay
12 McDonald Island
13 Tracy
14 Vernalis
15 Chowchilla
16 Gill Ranch
17 Trico
18 Semi-tropic
19 Buttonwillow
20 Johe Ranch
Source: California Division of Mines and Geology.
0
50
100
MILES

IRRIGATED LAND IN THE WEST

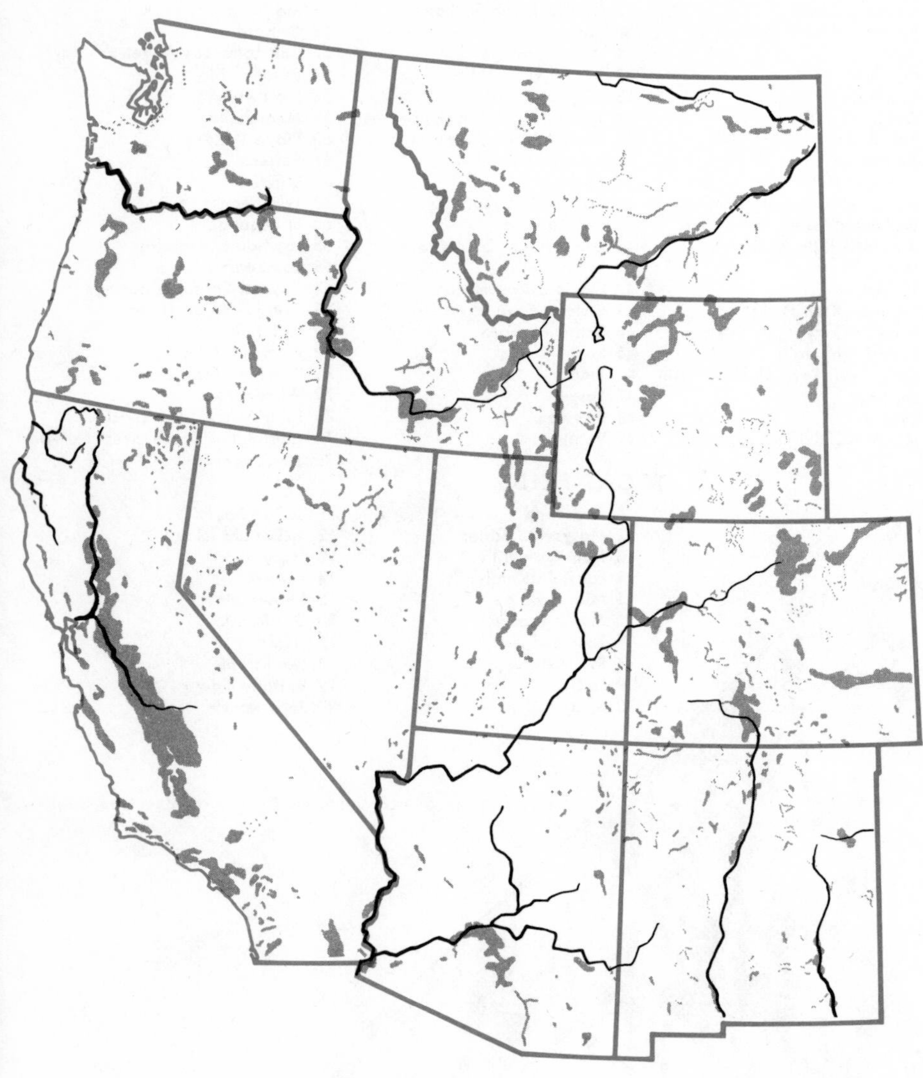

WATER RESOURCES

California's water resources represent one of her greatest assets and at times present her with some of her greatest problems. With the available water California has been able to develop eight million acres of irrigated land and build one of the world's largest urban complexes.

Water has always played a significant role in the life of the state. The location of Indian villages and Spanish settlements was predicated upon an adequate supply of water. Gold was first discovered along streams, and water was essential in the separation of gold from the gravel in which it was found.

The Spanish Padres diverted water from streams in the vicinity of the missions to irrigate their fields and to provide a domestic water supply. In the gold fields, ditches were dug to transport water to the mining areas. When mining declined in importance, these canals were converted to irrigate farms in the Central Valley. During this early period storage facilities were either very small or entirely lacking so that water had to be used when it was available.

However, it was not long before the need for adequate storage facilities to regulate the flow of water in irrigation canals was recognized. Several important storage reservoirs were constructed in southern California during the 1880's. Included were Bear Valley, Cuyamaca, Hemet and Sweetwater reservoirs. It was not until after the drought of the 1890's that major water storage facilities were constructed in other parts of the state.

The large metropolitan areas of the state were the first to feel the pressure of ever-increasing requirements for water and took the lead in the development of additional sources of water. The case of Los Angeles illustrates this point. When it found that local supplies of water were inadequate to care for its growing population, the city constructed an aqueduct 238 miles long from the Owens Valley and later extended it to Mono County.

The Owens Valley Aqueduct was followed by another project in which Los Angeles was joined by other communities in southern California. The Colorado Aqueduct was built in the 1930's after it became apparent that southern California needed additional water.

Water for agricultural and urban use in other parts of the state has been provided by various local, state and federal agencies. The Bureau of

COPCO #1
CLEAR LAKE RES.
TULE LAKE
GOOSE LAKE
BIG SAGE RES.
DWINNELL RES.
ALKALAI LAKE
TULE LAKE RES.
TRINITY RES.
LAKE BRITTON
SPOONER RES.
SHASTA RES.
EAGLE LAKE
LEWISTON RES.
SHASTA DAM
MC COY FLAT RES.
SWEASEY RES.
KESWICK RES.
WHISKEYTOWN RES.
HONEY LAKE
LAKE ALMANOR
MOUNTAIN MEADOWS RES.
Sacramento Valley Irrigation Canals
BUCKS LAKE
STONY GORGE RES.
OROVILLE RES.
BOCA RES.
LAKE PILLSBURY
BOWMAN RES.
COYOTE VALLEY RES.
EAST PARK RES.
FORDICE LAKE
ENGLEBRIGHT RES.
LAKE SPAULDING
CLEAR LAKE
LAKE TAHOE
FOLSOM RES.
UNION VALLEY RES.
FOLSOM DAM
SLY PARK RES.
TOPAZ LAKE
LAKE BERRYESSA
LAKE NATOMA
NASHVILLE RES.
SALT SPRINGS RES.
PARDEE RES.
SPICER MEADOW RES.
Delta-Cross Channel
HOGAN RES.
BRIDGEPORT RES.
LAKE LLOYD
KENT RES.
FARMINGTON RES.
MELONES RES.
LAKE ELEANOR
MONO LAKE
WOODWARD RES.
TULLOCH RES.
HETCH HETCHY RES.
GRANT LAKE
SAN PABLO RES.
SAN LEANDRO RES.
Contra Costa Canal
Aqueduct
DON PEDRO RES.
SAN ANDREAS LAKE
CRYSTAL SPRINGS RES.
Hetch-Hetchy
MODESTO RES.
CROWLEY LAKE
EXCHEQUER RES.
TURLOCK RES.
CALAVERAS RES.
BASS LAKE
EDISON RES.
ANDERSON RES.
HIDDEN RES.
HUNTINGTON LAKE
Delta-Mendota Canal
Madera Canal
SHAVER LAKE
TINEMAHA RES.
SAN LUIS RES.
FRIANT DAM
MILLERTON LAKE
WISHON RES.
PINE FLAT RES.
MENDOTA POOL
Friant-Kern Canal
OWENS LAKE
TERMINUS RES.
HAIWEE RES.
TULARE LAKE
SUCCESS RES.
California Aqueduct
ISABELLA RES.
Aqueduct
NACIMIENTO RES.
SALINAS RES.
BUENA VISTA LAKE
Los Angeles
TWITCHELL RES.
GIBRALTAR RES.
MATILIJA RES.
LAKE PIRU
BOUQUET RES.
LAKE HAVASU
PARKER DAM
CACHUMA RES.
LAKE CASITAS
CHATSWORTH RES.
SAN FERNANDO RES.
HANSEN DAM
BIG BEAR LAKE
SEPULVEDA DAM
LAKE ARROWHEAD
LAKE MATHEWS
Colorado River Aqueduct
SANTIAGO RES.
HAYFIELD RES.
LAKE ELSINORE
VAIL RES.
Coachella Branch
HENSHAW RES.
SALTON SEA
IMPERIAL DAM
HODGES RES.
CUYAMACA LAKE
SAN VICENTE RES.
EL CAPITAN RES.
All-American Canal
SWEETWATER RES.
MORENA LAKE
OTAY RES.
BARRETT LAKE
0 50 100
MILES

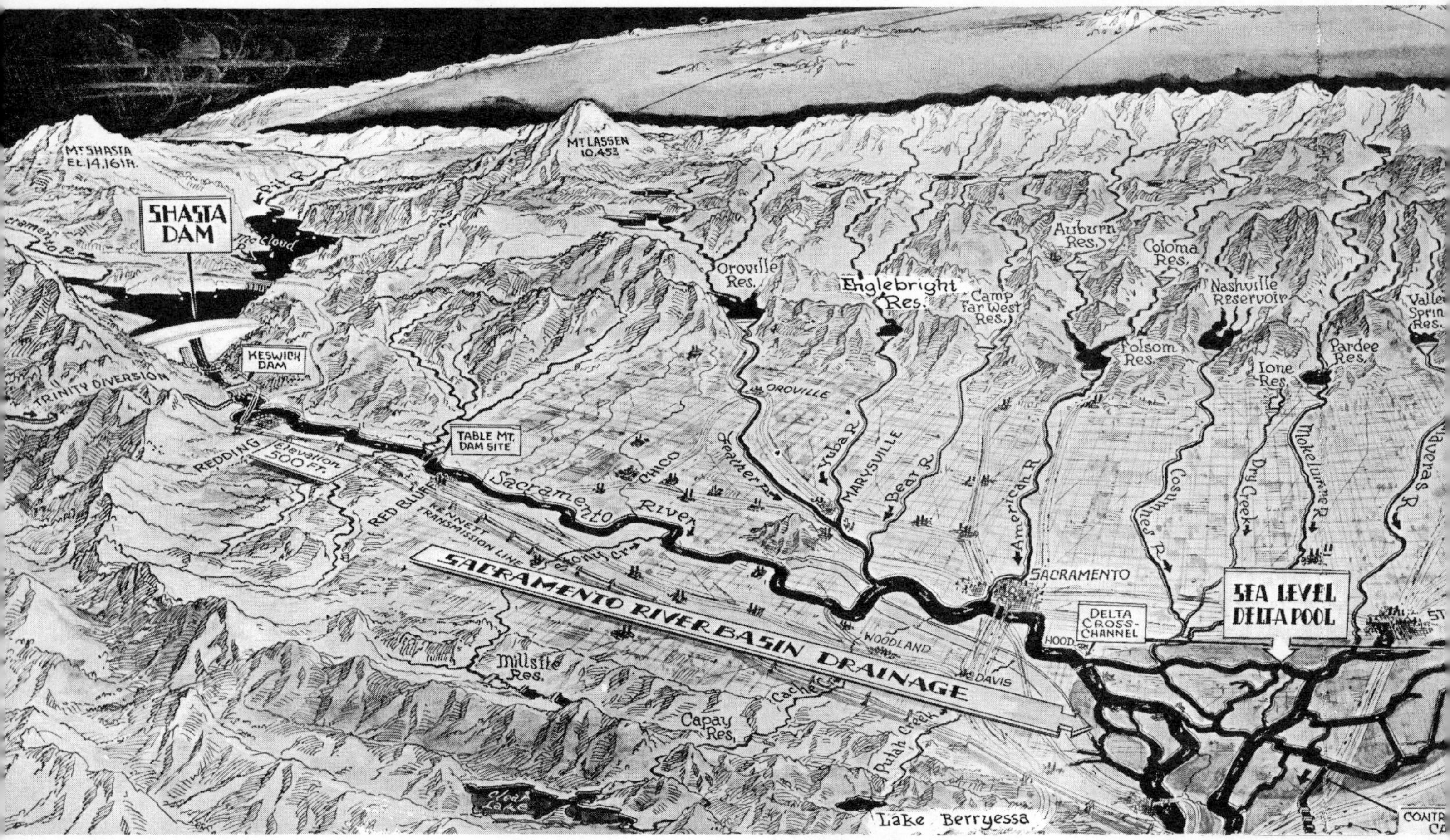

COLORADO RIVER BASIN

Friant Dam and the Friant-Kern Canal.

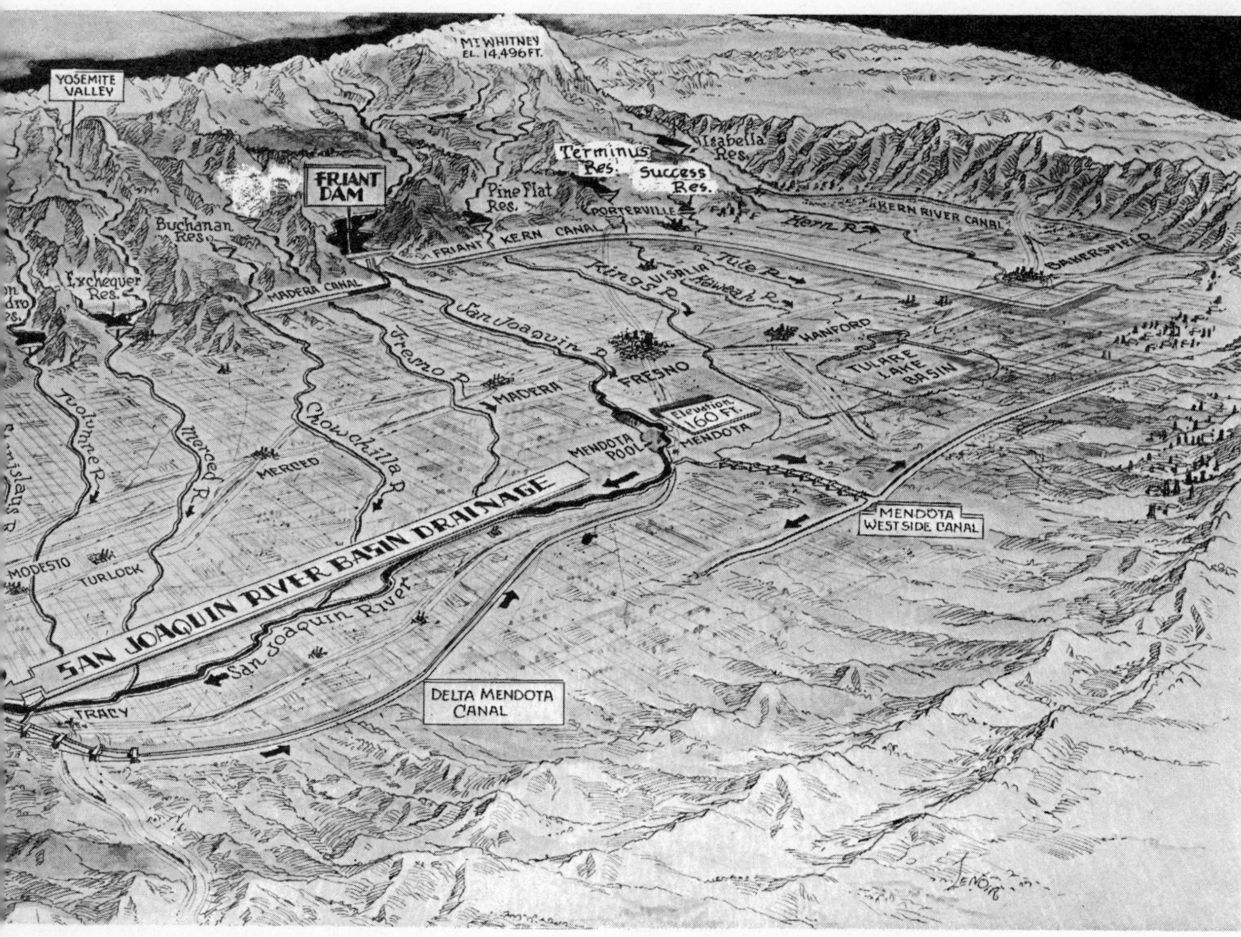

Reclamation has supervised construction of the Central Valley Project and of the various facilities along the Colorado River including the All-American Canal which brings irrigation water to the Imperial and Coachella valleys.

Oroville Dam and the California Aqueduct are the principal elements of the State Water Project completed in 1972. Water from Northern California is transported southward to the San Joaquin and Antelope valleys and to Southern California. The search for additional sources of water to accommodate future growth has included discussion of the possibility of tapping the Snake or Columbia rivers and the desalination of ocean water. Others have suggested that the urban areas of the Southwest are already too crowded and that additional population growth should be discouraged.

Crescent City
Lava Beds Nat'l Mon.
Alturas
Dunsmuir
EUREKA
REDDING
LASSEN VOLCANIC NAT'L PARK
Susanville
Red Bluff
Corning
Quincy
Chico
Willits
Willows
Oroville
Truckee
Marysville
Auburn
Santa Rosa
SACRAMENTO
Napa
Lodi
STOCKTON
Muir Woods Nat'l Mon.
SAN FRANCISCO
YOSEMITE NAT'L PARK
Devils Postpile Nat'l Mon.
Livermore
MODESTO
Palo Alto
SAN JOSE
Merced
Bishop
Santa Cruz
Madera
KINGS CANYON NAT'L PARK
FRESNO
SEQUOIA NAT'L PARK
Death Valley Nat'l Mon.
Salinas
Monterey
Pinnacles Nat'l Mon.
King City
Visalia
Hanford
Tulare
Coalinga
Porterville
Delano
Paso Robles
San Luis Obispo
BAKERSFIELD
Taft
Maricopa
Santa Maria
Lancaster
Barstow
Needles
Palmdale
Lompoc
Santa Barbara
Ventura
Oxnard
LOS ANGELES
Channel Islands Nat'l Monument
SAN BERNARDINO
Riverside
Joshua Tree Nat'l Mon.
Palm Springs
LONG BEACH
Santa Ana
Indio
Blythe
Oceanside
Brawley
Cabrillo Nat'l Mon.
El Centro
Calexico
SAN DIEGO
1. Silver Strand Beach
2. Cuyamaca Rancho
3. Palomar Mountain
4. Borrego
5. Anza Desert
6. Salton Sea
7. Mission Bay
8. Torrey Pines Beach
9. Cardiff Beach
10. San Elijo Beach
11. Moonlight Beach
12. Ponto Beach
13. La Costa Beach
14. Carlsbad Beach
15. San Clemente Beach
16. Doheny Beach
17. Mt. San Jacinto
18. Corona Del Mar Beach
19. Huntington Beach
20. Alamitos Beach
21. Redondo Beach
22. Manhattan Beach
23. Dockweiler Beach
24. Santa Monica Beach
25. Will Rogers Beach
26. Las Tunas Beach
27. Malibu Lagoon Beach
28. Leo Carrillo Beach
29. Los Angeles State Arboretum
30. Placerita Canyon
31. Mitchell Caverns
32. San Buenaventura Beach
33. Arroyo Burro Beach
34. Goleta Beach
35. El Capitan Beach
36. Refugio Beach
37. Gaviota Beach
38. Will Rogers
39. Point Sal Beach
40. Pismo Beach
41. Avila Beach
42. Morro Beach
43. Morro Strand Beach
44. Cayucos Beach
45. San Simeon Beach
46. John Little
47. Pfeiffer - Big Sur
48. Tule Elk Reserve
49. Kern River
50. Montgomery Memorial
51. Carpinteria Beach
52. Emma Wood Beach
53. Joshua Trees
54. Fort Tejon
55. Hearst - San Simeon
56. La Purisima
57. Los Encinos
58. Lummis Home
59. Pio Pico
60. San Pasqual Battlefield
61. Bolsa Chica
62. Lake Elsinore
100. Point Lobos Reserve
101. Carmel River Beach
102. Asilomar Beach
103. Fremont Peak
104. Zmudowski Beach
105. Sunset Beach
106. Manresa Beach
107. Seacliff Beach
108. New Brighton Beach
109. Capitola Beach
110. Natural Bridges Beach
111. Henry Cowell Redwoods
112. Big Basin Redwoods
113. Portola
114. Fremont Ford
115. George J. Hatfield
116. McConnell
117. Turlock Lake
118. Caswell
119. Brannan Island
120. Mount Diablo
121. Angel Island
122. James D. Phelan Beach
123. Baker's Beach
124. Stinson Beach
125. Mount Tamalpais
126. Samuel P. Taylor
127. Tomales Bay
128. Sonoma Coast
129. Armstrong Redwoods
130. Robert Louis Stevenson Memorial
131. Kruse Rhododendron Reserve
132. Clear Lake
133. Mailliard Redwoods
134. Indian Creek
135. Paul M. Dimmick Memorial
136. Montgomery Woods
137. Van Damme Beach
138. Russian Gulch
139. MacKerricher Beach
140. Westport-Union Landing Beach
141. Admiral William H. Standley
142. Edward R. Hickey
143. Richardson Grove
144. Stephens Grove
145. Williams Grove
146. Burlington
147. Dyerville
148. Bull Creek
149. Grizzly Creek Redwoods
150. Azalea Reserve
151. Little River Beach
152. Trinidad Beach
153. Patrick's Point
154. Dry Lagoon Beach
155. Prairie Creek Redwoods
156. Del Norte Creek
157. Jedediah Smith Redwoods
158. Pelican Beach
159. Castle Crags
160. McArthur-Burney Falls Memorial
161. Curry-Bidwell Bar
162. Donner Memorial
163. Tahoe
164. D. L. Bliss
165. Emerald Bay
166. James W. Marshall
167. Calaveras Big Trees
168. Knowland State Park & Arboretum
169. Butano
170. San Mateo Beaches
171. Twin Lakes Beach
172. Colusa - Sacramento River
173. Folsom Lake
174. Columbia Historic
175. Millerton Lake
176. Benicia
177. Fort Humboldt
178. Fort Ross
179. Monterey SH Monuments
180. Petaluma
181. San Juan Bautista
182. Shasta
183. State Indian Museum
184. Sutter's Fort
185. Weaverville Joss House
186. Benbow Lake
187. Plumas - Eureka
188. Wm. B. Ide Adobe
189. Woodson Bridge
190. Ano Nuevo Beach
191. Franks Tract
192. Grover Hot Springs
193. Henry W. Coe

INDUSTRIAL GROWTH

The events associated with World War II revolutionized the California economy, which had been based upon agriculture, mining, forestry and tourism, and established manufacturing as the most important segment of the economy.

Prior to that time it was said that Californians took in each other's wash to support themselves, and, while this was not entirely accurate, a disproportionate share of the population was engaged in transporting and selling goods and in providing services for others.

During the Spanish and Mexican period most manufactured goods were imported, and Yankee Traders supplied the needs of the Californias. In the Gold Rush Period the manufacture of mining equipment and of consumers' goods such as clothing, food and beverages briefly flourished as San Francisco became the ninth largest manufacturing center in the country.

The completion of the transcontinental railroad brought an influx of manufactured goods from the East, and local plants which could not compete were forced to close. Californians again were dependent on outside sources for most of their manufactured goods.

During the remainder of the 19th century small industries supplying local markets and processing agricultural products represented the only significant contribution of the industrial segment of the economy. In 1900 the state had about 2 per cent of the U.S. manufacturing output whereas now the figure is almost 9 per cent.

With a population consisting largely of single men and with a great number of visitors and newcomers to the state the demand for services of all kinds was rather great. The intensive, irrigated agriculture which had begun to develop had heavy labor demands.

The first decade of the 20th century brought a new development of southern California — the movie industry. From a modest beginning it grew rapidly to become an important factor in the California economy. Industrial firms in other

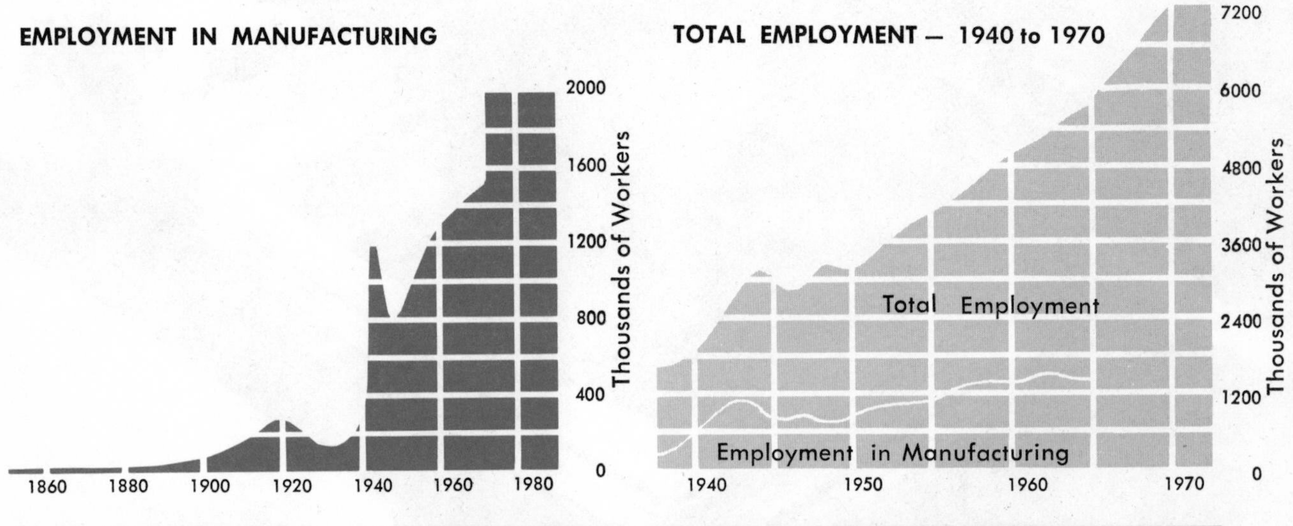

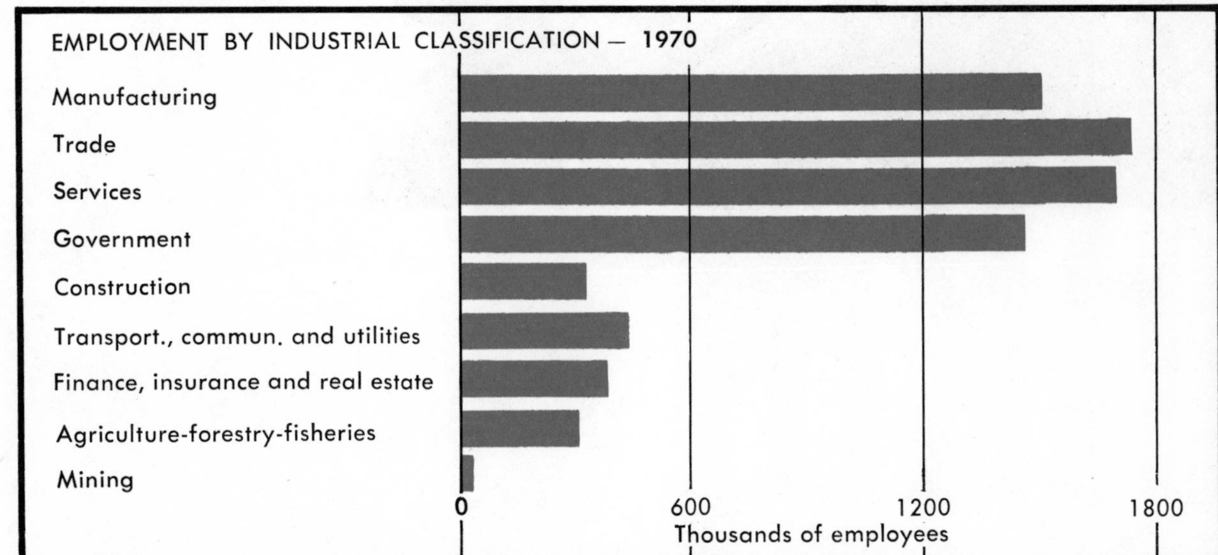

fields were added slowly as the years moved forward. Automobile assembly plants, aircraft plants, apparel factories and iron and steel mills became important. Then, war-born industrialization changed the entire economic status of the state. Employment in shipbuilding and in aircraft and ordnance factories expanded the industrial base markedly. Employment in manufacturing rose from 260,800 in 1938 to 445,000 in 1941 and 722,000 by 1947.

With the close of World War II everyone assumed that the economy of California would return to its pre-war status and that the shipyard workers and aircraft assemblers would return to their old homes and their old jobs. Instead, most of them remained in California and were absorbed into the labor force as local industries expanded to supply the enlarged West Coast market.

Since the late 40's the Cold War has increased federal spending and enlarged employment in the aerospace industries. Companies like North American, Lockheed, Litton, Aerojet-General, General Dynamics, Hughes and Northrup employ about half a million people. Over half of these work in Los Angeles County.

Expansion of the total West Coast market has meant the constant growth of consumer-oriented industries to meet the demands of new arrivals in the area. Construction and furniture industries satisfy needs for new homes while expanded facilities for processing the products of California's farms provide foods and beverages. So long as the aerospace firms are able to procure government contracts the "new" economy of the state will continue to grow and prosper.

EMPLOYMENT IN MANUFACTURING

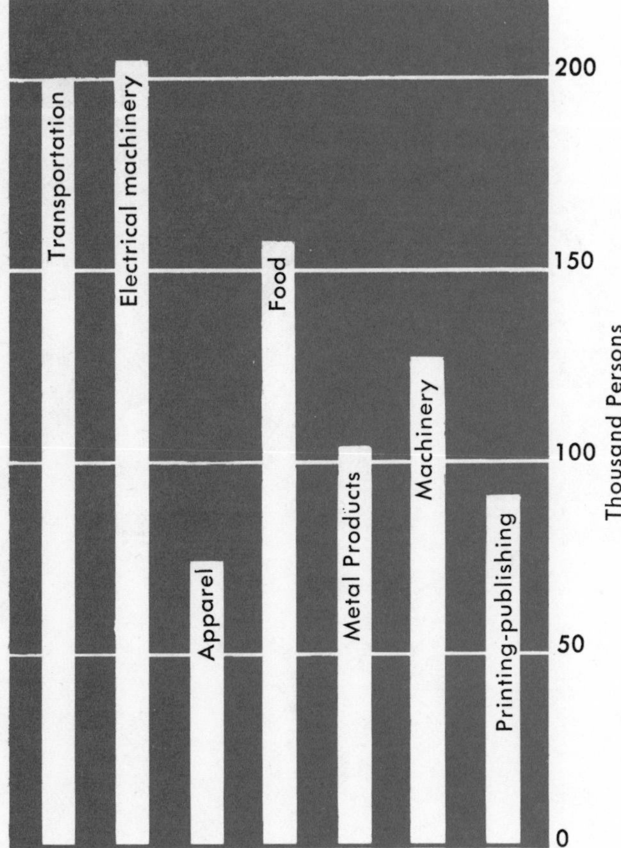

NUMBER OF INDUSTRIAL EMPLOYEES

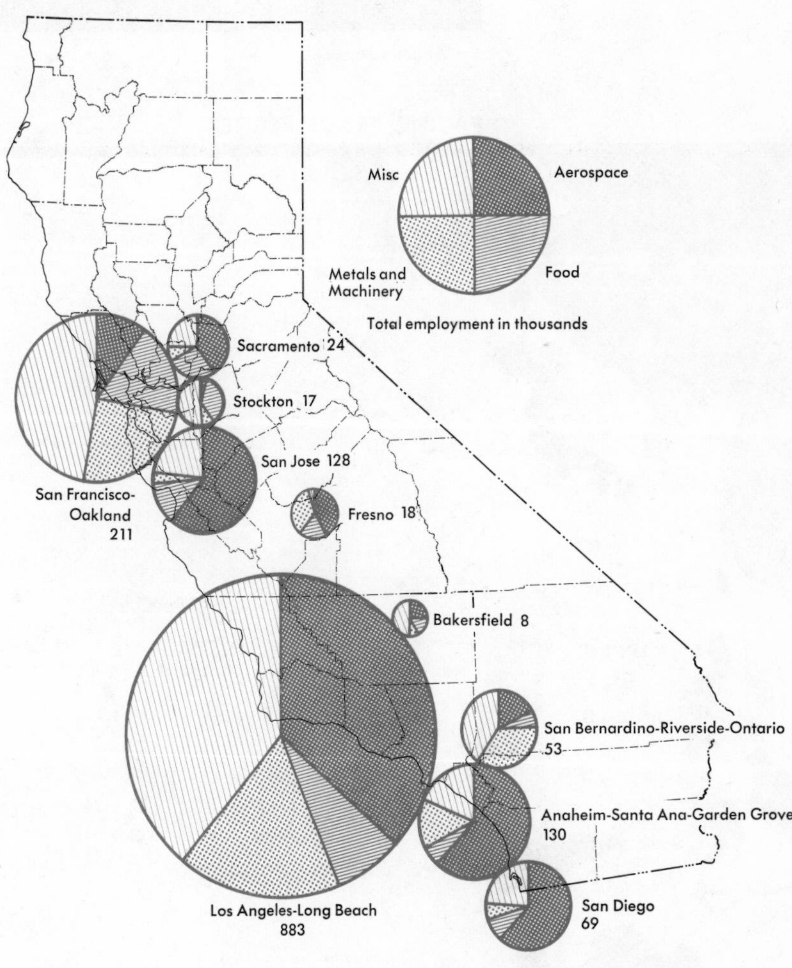

Los Angeles Harbor.

PACIFIC TRADE ROUTES

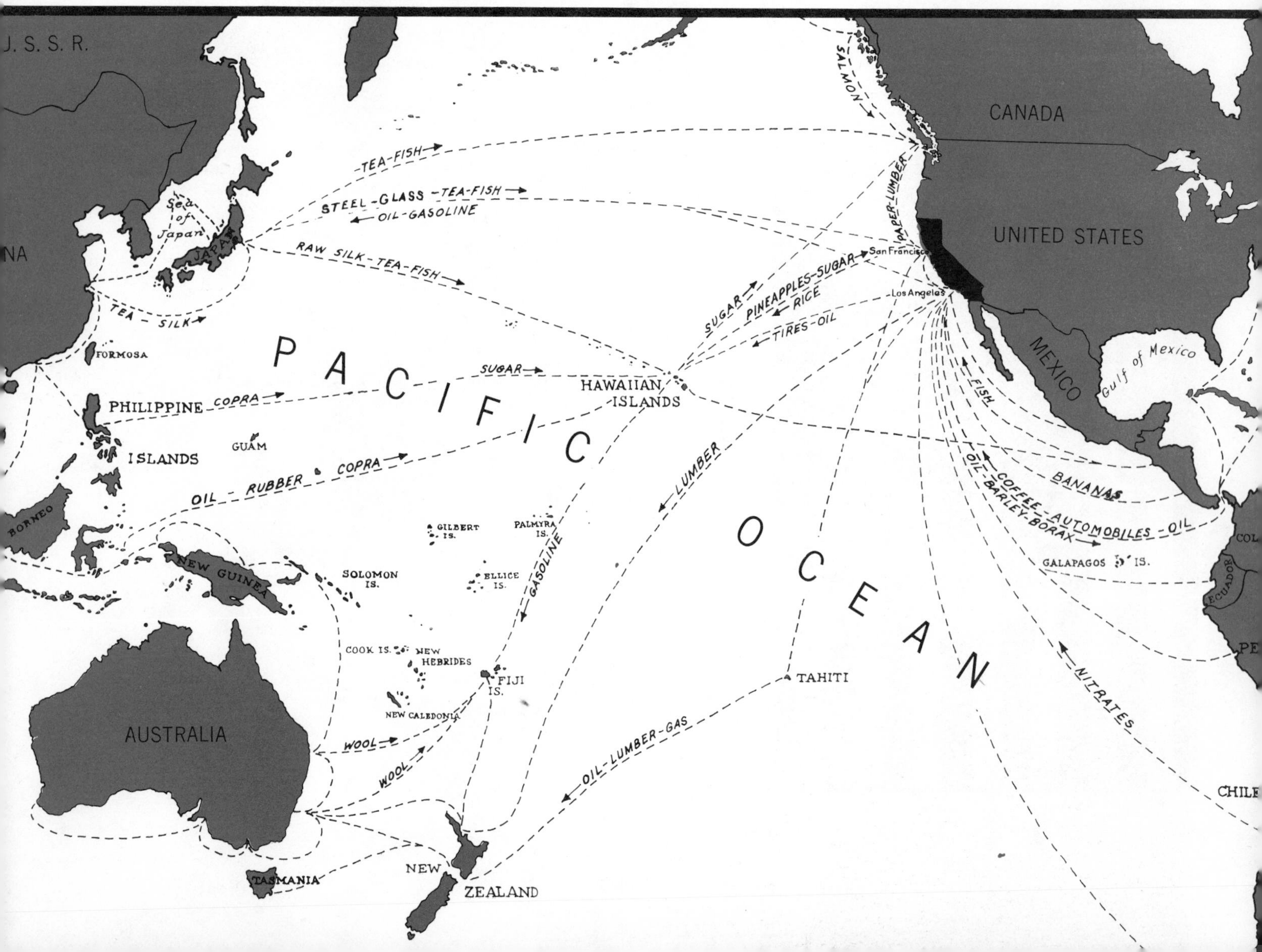

J. S. S. R.
CANADA
UNITED STATES
MEXICO
Gulf of Mexico
NA
Sea of Japan
JAPAN
FORMOSA
PHILIPPINE ISLANDS
GUAM
BORNEO
NEW GUINEA
SOLOMON IS.
GILBERT IS.
PALMYRA IS.
ELLICE IS.
COOK IS.
NEW HEBRIDES
NEW CALEDONIA
FIJI IS.
TAHITI
AUSTRALIA
TASMANIA
NEW ZEALAND
HAWAIIAN ISLANDS
San Francisco
Los Angeles
GALAPAGOS IS.
ECUADOR
COL
PE
CHILE
PACIFIC OCEAN
SALMON
TEA-FISH
STEEL-GLASS -TEA-FISH
OIL-GASOLINE
RAW SILK-TEA-FISH
PAPER-LUMBER
TEA - SILK
SUGAR
PINEAPPLES-SUGAR
RICE
TIRES-OIL
SUGAR
COPRA
OIL - RUBBER - COPRA
LUMBER
GASOLINE
FISH
BANANAS
COFFEE-AUTOMOBILES-OIL
OIL-BARLEY-BORAX
NITRATES
WOOL
WOOL
OIL-LUMBER-GAS

FOREIGN TRADE

From the time of the arrival of the Spaniards to the present, ships from all parts of the earth have come to California to trade. In Spanish and Mexican days furs, hides and tallow were exchanged for manufactured goods from England and from northeastern United States.

Ships brought a large number of people to San Francisco during the Gold Rush Period, and it became California's chief port and most important city during the 19th century. Goods and people from all parts of the world entered and left the Bay Region.

In the 20th century a number of factors have contributed to the increased significance of southern California ports. The growth of population in the area, the development of industry, the construction of excellent harbor facilities, and the completion of the Panama Canal have all been important in accounting for the enlarged flow of ocean traffic through the ports of Los Angeles, Long Beach and San Diego.

Today, San Francisco continues to be the principal port for the movement of passenger traffic and of general cargo. It receives and processes basic raw materials from countries around the rim of the Pacific before forwarding them to other parts of the country. The ports of Los Angeles and Long Beach handle a greater volume of cargo than does San Francisco, but the bulk of it is crude and refined petroleum products. However, cotton for export and lumber to build the suburbs are important general cargo items. The harbor at San Diego functions chiefly as the home port of the Pacific fleet although small quantities of general cargo flow in and out of her port facilities.

San Diego Harbor.

San Francisco Harbor.

OCEAN TRAFFIC—1969 (Tonnage)

	Foreign		Domestic	
	Imports	Exports	Imports	Exports
Crescent City	—	—	218,967	28,448
Humboldt Bay	—	760,612	464,495	35,199
Carquinez Strait	1,288,475	284,353	5,553,317	1,631,494
San Pablo Bay	1,286,154	22,492	1,689,079	1,593,674
Richmond	1,640,358	604,692	7,988,627	3,703,518
Oakland	756,084	1,258,665	656,560	965,446
Redwood City	171,511	430,059	84,831	584,567
San Francisco	1,245,042	1,004,216	28,507	130,763
Suisun Bay	79,888	55,214	—	—
Stockton	102,934	788,496	139,379	37,079
Sacramento	4,581	1,115,710	373,549	5,364,125
Los Angeles	5,786,371	4,222,415	5,661,945	5,996,445
Long Beach	4,797,381	6,180,710	4,414,188	3,479,096
San Diego	272,449	509,390	746,679	1,083

Los Angeles International Airport.

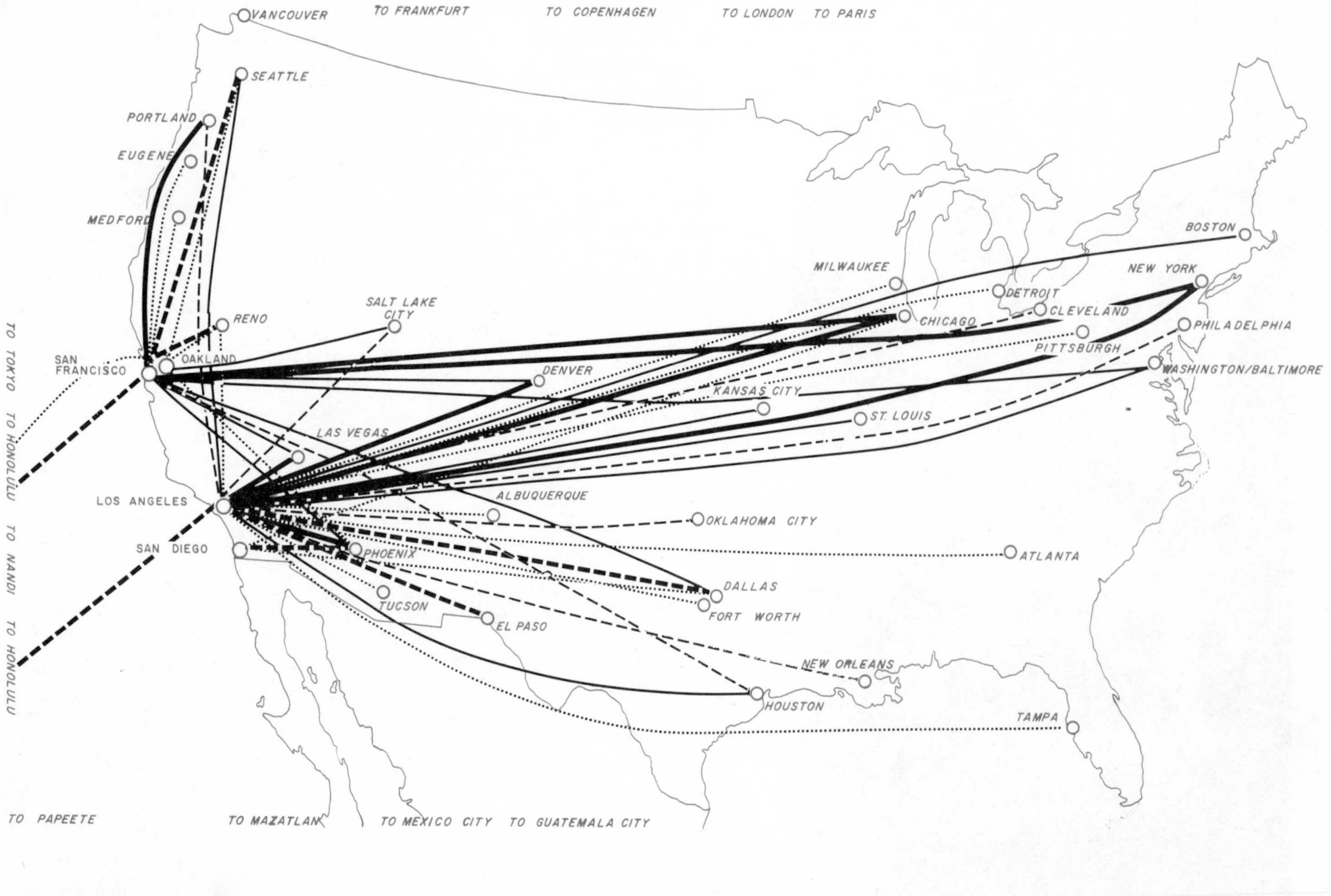

VANCOUVER
TO FRANKFURT
TO COPENHAGEN
TO LONDON
TO PARIS
SEATTLE
PORTLAND
EUGENE
MEDFORD
BOSTON
MILWAUKEE
NEW YORK
DETROIT
SALT LAKE CITY
CLEVELAND
RENO
CHICAGO
PHILADELPHIA
TO TOKYO
SAN FRANCISCO
OAKLAND
PITTSBURGH
DENVER
WASHINGTON/BALTIMORE
TO HONOLULU
KANSAS CITY
ST. LOUIS
LAS VEGAS
LOS ANGELES
ALBUQUERQUE
TO NANDI
OKLAHOMA CITY
SAN DIEGO
PHOENIX
ATLANTA
DALLAS
TUCSON
FORT WORTH
EL PASO
TO HONOLULU
NEW ORLEANS
HOUSTON
TAMPA
TO PAPEETE
TO MAZATLAN
TO MEXICO CITY
TO GUATEMALA CITY

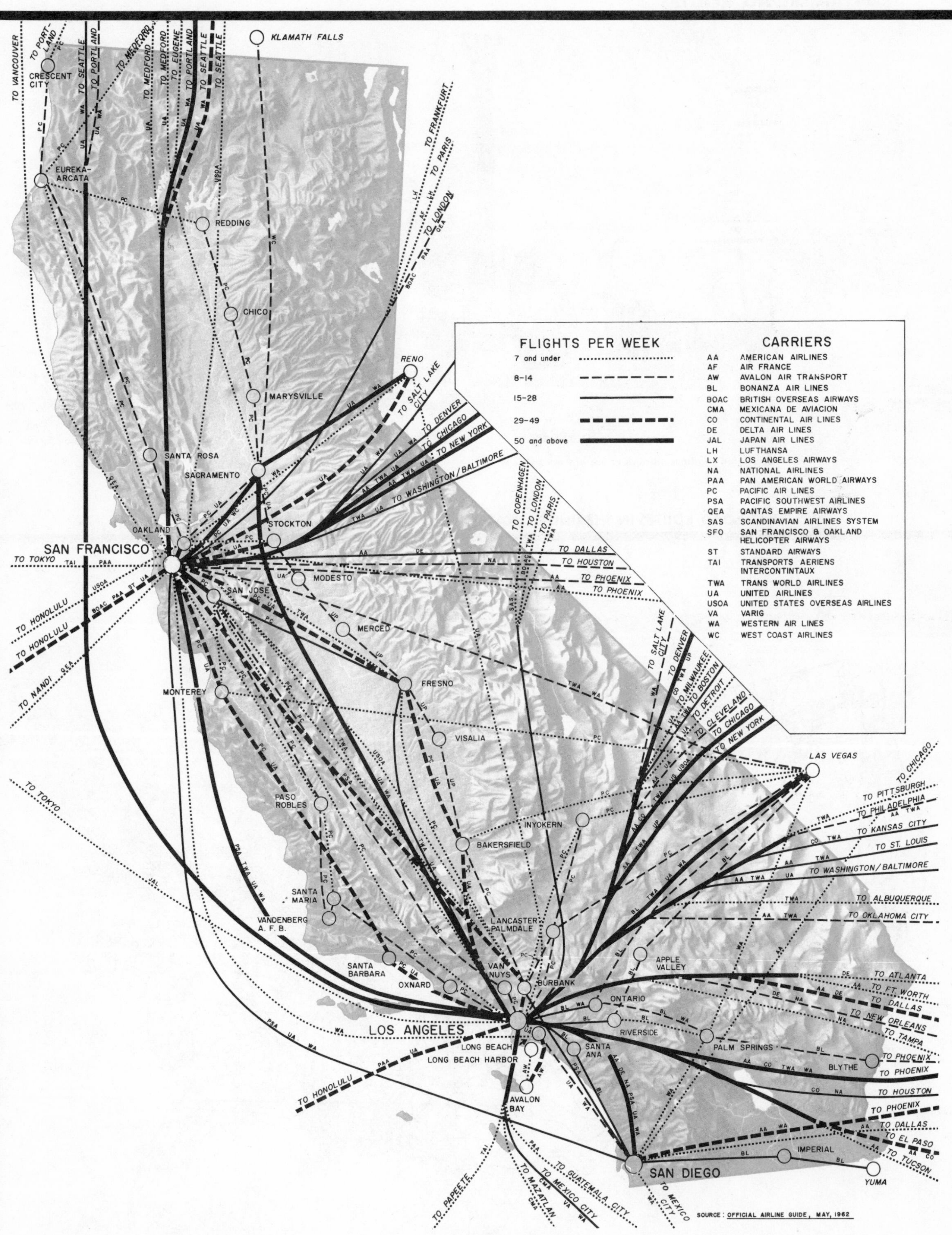

FLIGHTS PER WEEK
7 and under
8–14
15–28
29–49
50 and above
CARRIERS
AA AMERICAN AIRLINES
AF AIR FRANCE
AW AVALON AIR TRANSPORT
BL BONANZA AIR LINES
BOAC BRITISH OVERSEAS AIRWAYS
CMA MEXICANA DE AVIACION
CO CONTINENTAL AIR LINES
DE DELTA AIR LINES
JAL JAPAN AIR LINES
LH LUFTHANSA
LX LOS ANGELES AIRWAYS
NA NATIONAL AIRLINES
PAA PAN AMERICAN WORLD AIRWAYS
PC PACIFIC AIR LINES
PSA PACIFIC SOUTHWEST AIRLINES
QEA QANTAS EMPIRE AIRWAYS
SAS SCANDINAVIAN AIRLINES SYSTEM
SFO SAN FRANCISCO & OAKLAND HELICOPTER AIRWAYS
ST STANDARD AIRWAYS
TAI TRANSPORTS AERIENS INTERCONTINTAUX
TWA TRANS WORLD AIRLINES
UA UNITED AIRLINES
USOA UNITED STATES OVERSEAS AIRLINES
VA VARIG
WA WESTERN AIR LINES
WC WEST COAST AIRLINES
KLAMATH FALLS
CRESCENT CITY
EUREKA-ARCATA
REDDING
CHICO
RENO
MARYSVILLE
SANTA ROSA
SACRAMENTO
OAKLAND
STOCKTON
SAN FRANCISCO
MODESTO
SAN JOSE
MERCED
FRESNO
MONTEREY
VISALIA
LAS VEGAS
PASO ROBLES
INYOKERN
BAKERSFIELD
SANTA MARIA
VANDENBERG A. F. B.
LANCASTER PALMDALE
APPLE VALLEY
SANTA BARBARA
OXNARD
VAN NUYS
BURBANK
ONTARIO
LOS ANGELES
RIVERSIDE
LONG BEACH
LONG BEACH HARBOR
SANTA ANA
PALM SPRINGS
BLYTHE
AVALON BAY
IMPERIAL
SAN DIEGO
YUMA
TO VANCOUVER
TO PORTLAND
TO SEATTLE
TO MEDFORD
TO EUGENE
TO FRANKFURT
TO PARIS
TO LONDON
TO SALT LAKE CITY
TO DENVER
TO CHICAGO
TO NEW YORK
TO WASHINGTON/BALTIMORE
TO COPENHAGEN
TO DALLAS
TO HOUSTON
TO PHOENIX
TO TOKYO
TO HONOLULU
TO NANDI
TO MILWAUKEE
TO BOSTON
TO DETROIT
TO CLEVELAND
TO PITTSBURGH
TO PHILADELPHIA
TO KANSAS CITY
TO ST. LOUIS
TO ALBUQUERQUE
TO OKLAHOMA CITY
TO ATLANTA
TO FT. WORTH
TO NEW ORLEANS
TO TAMPA
TO EL PASO
TO TUCSON
TO PAPEETE
TO MAZATLAN
TO MEXICO CITY
TO GUATEMALA CITY
SOURCE: OFFICIAL AIRLINE GUIDE, MAY, 1962

California products enroute to eastern markets.

RAILROAD ROUTES IN THE UNITED STATES

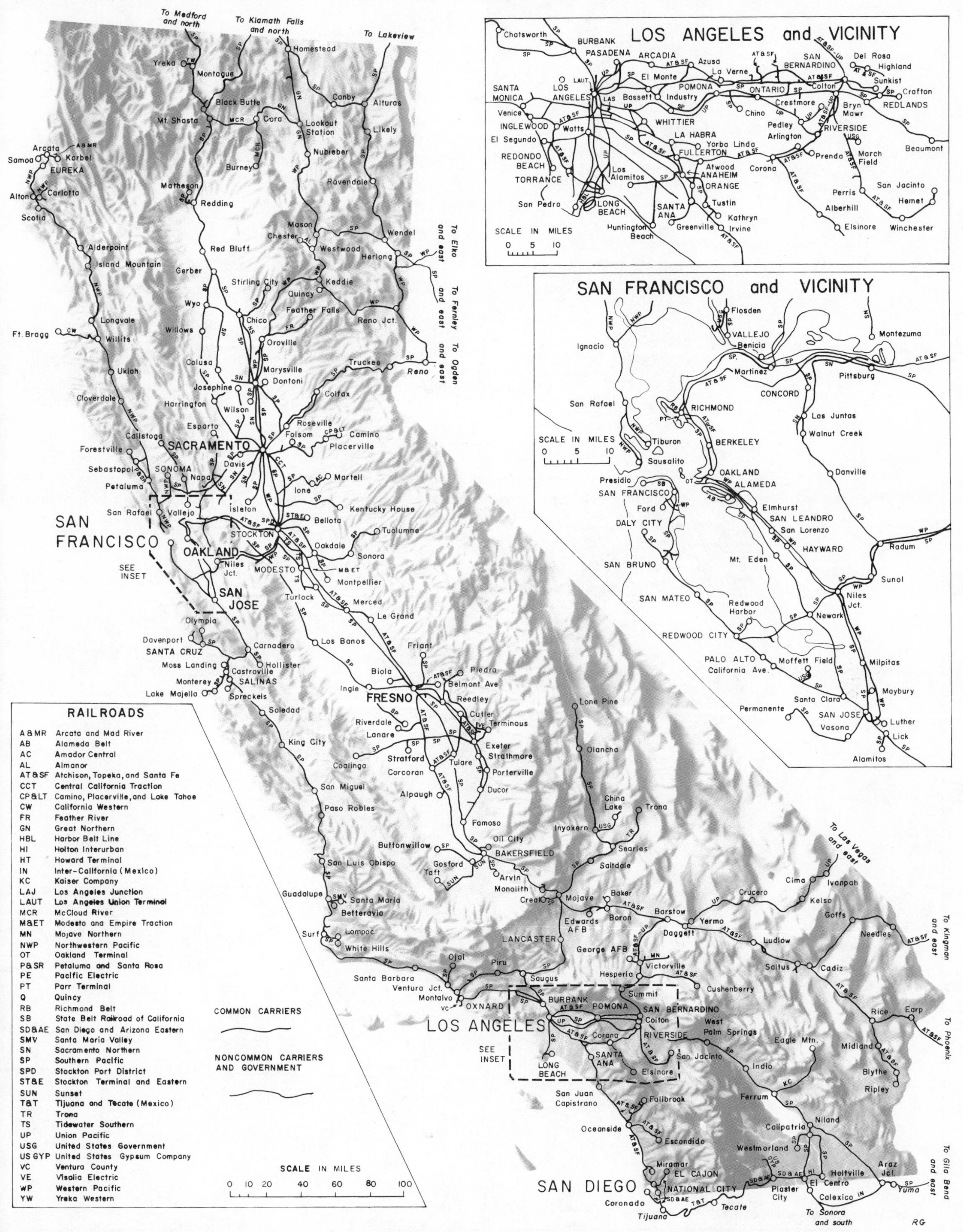
LOS ANGELES and VICINITY
SCALE IN MILES
0 5 10
SAN FRANCISCO and VICINITY
SCALE IN MILES
0 5 10
SAN FRANCISCO
SEE INSET
OAKLAND
SAN JOSE
SACRAMENTO
STOCKTON
FRESNO
BAKERSFIELD
LOS ANGELES
SEE INSET
SAN DIEGO
To Medford and north
To Klamath Falls and north
To Lakeview
To Elko and east
To Fernley and east
To Ogden and east
To Las Vegas and east
To Kingman and east
To Phoenix
To Gila Bend and east
To Sonora and south
RAILROADS
A&MR Arcata and Mad River
AB Alameda Belt
AC Amador Central
AL Almanor
AT&SF Atchison, Topeka, and Santa Fe
CCT Central California Traction
CP< Camino, Placerville, and Lake Tahoe
CW California Western
FR Feather River
GN Great Northern
HBL Harbor Belt Line
HI Holton Interurban
HT Howard Terminal
IN Inter-California (Mexico)
KC Kaiser Company
LAJ Los Angeles Junction
LAUT Los Angeles Union Terminal
MCR McCloud River
M&ET Modesto and Empire Traction
MN Mojave Northern
NWP Northwestern Pacific
OT Oakland Terminal
P&SR Petaluma and Santa Rosa
PE Pacific Electric
PT Parr Terminal
Q Quincy
RB Richmond Belt
SB State Belt Railroad of California
SD&AE San Diego and Arizona Eastern
SMV Santa Maria Valley
SN Sacramento Northern
SP Southern Pacific
SPD Stockton Port District
ST&E Stockton Terminal and Eastern
SUN Sunset
T&T Tijuana and Tecate (Mexico)
TR Trona
TS Tidewater Southern
UP Union Pacific
USG United States Government
US GYP United States Gypsum Company
VC Ventura County
VE Visalia Electric
WP Western Pacific
YW Yreka Western
COMMON CARRIERS
NONCOMMON CARRIERS AND GOVERNMENT
SCALE IN MILES
0 10 20 40 60 80 100
RG

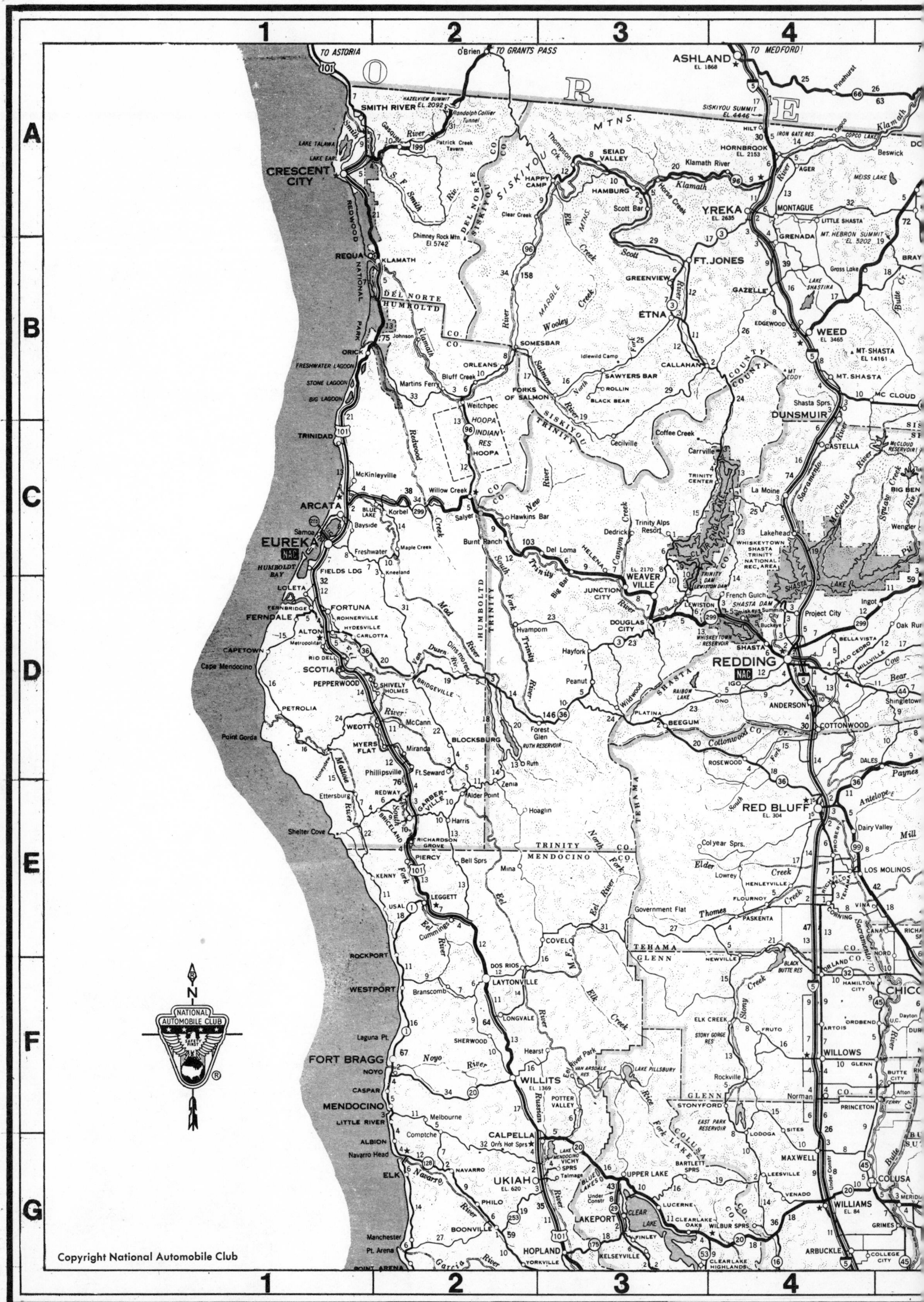

TO ASTORIA
TO GRANTS PASS
TO MEDFORD
ASHLAND
SMITH RIVER
CRESCENT CITY
YREKA
FT. JONES
ETNA
WEED
MT. SHASTA
DUNSMUIR
REDWOOD NATIONAL PARK
ORICK
TRINIDAD
ARCATA
EUREKA
HUMBOLDT BAY
FORTUNA
FERNDALE
SCOTIA
WEAVERVILLE
REDDING
ANDERSON
COTTONWOOD
RED BLUFF
CORNING
ORLAND
CHICO
WILLOWS
WILLITS
FORT BRAGG
MENDOCINO
UKIAH
LAKEPORT
CLEAR LAKE
WILLIAMS
COLUSA
HOPLAND
SISKIYOU MTNS.
NATIONAL AUTOMOBILE CLUB
Copyright National Automobile Club

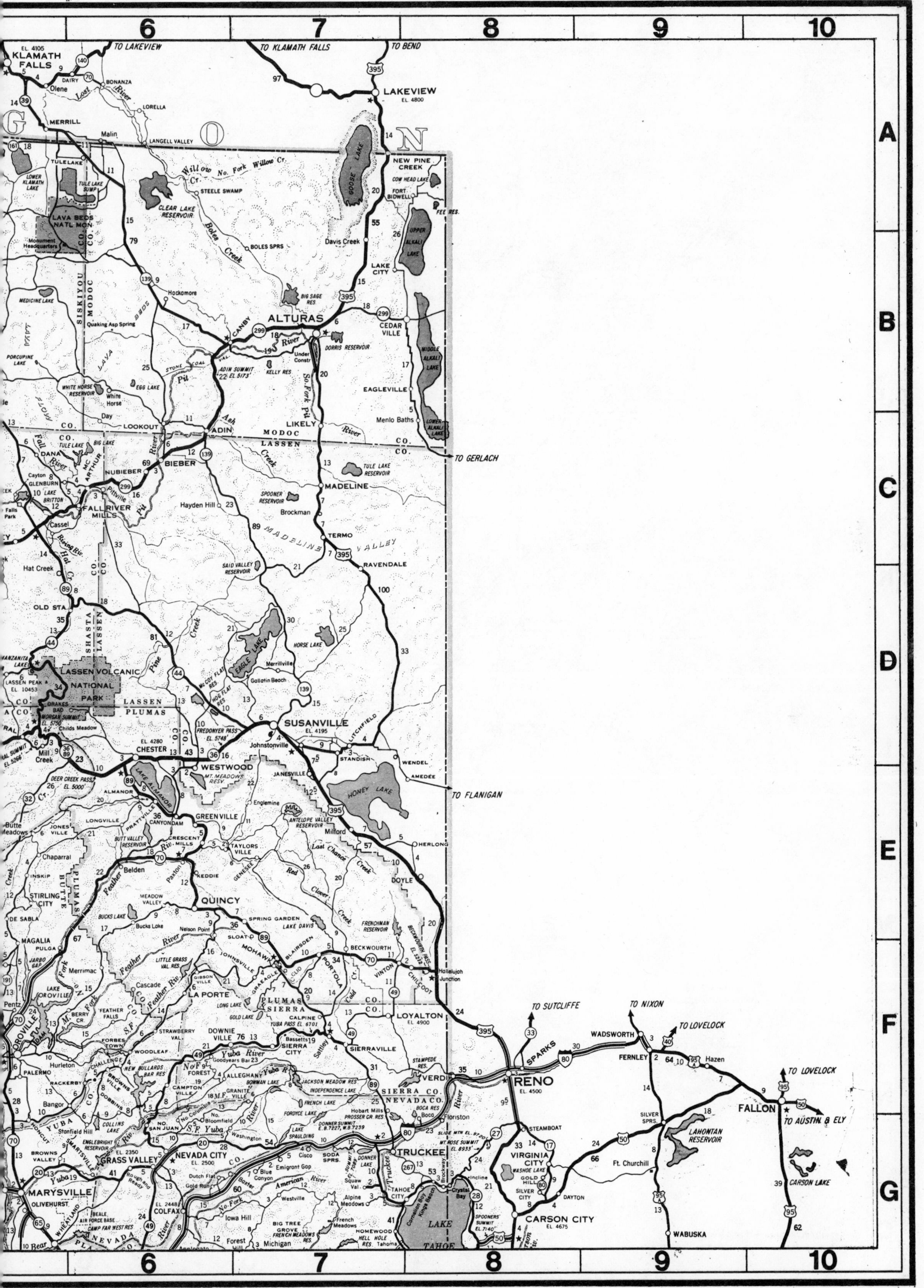

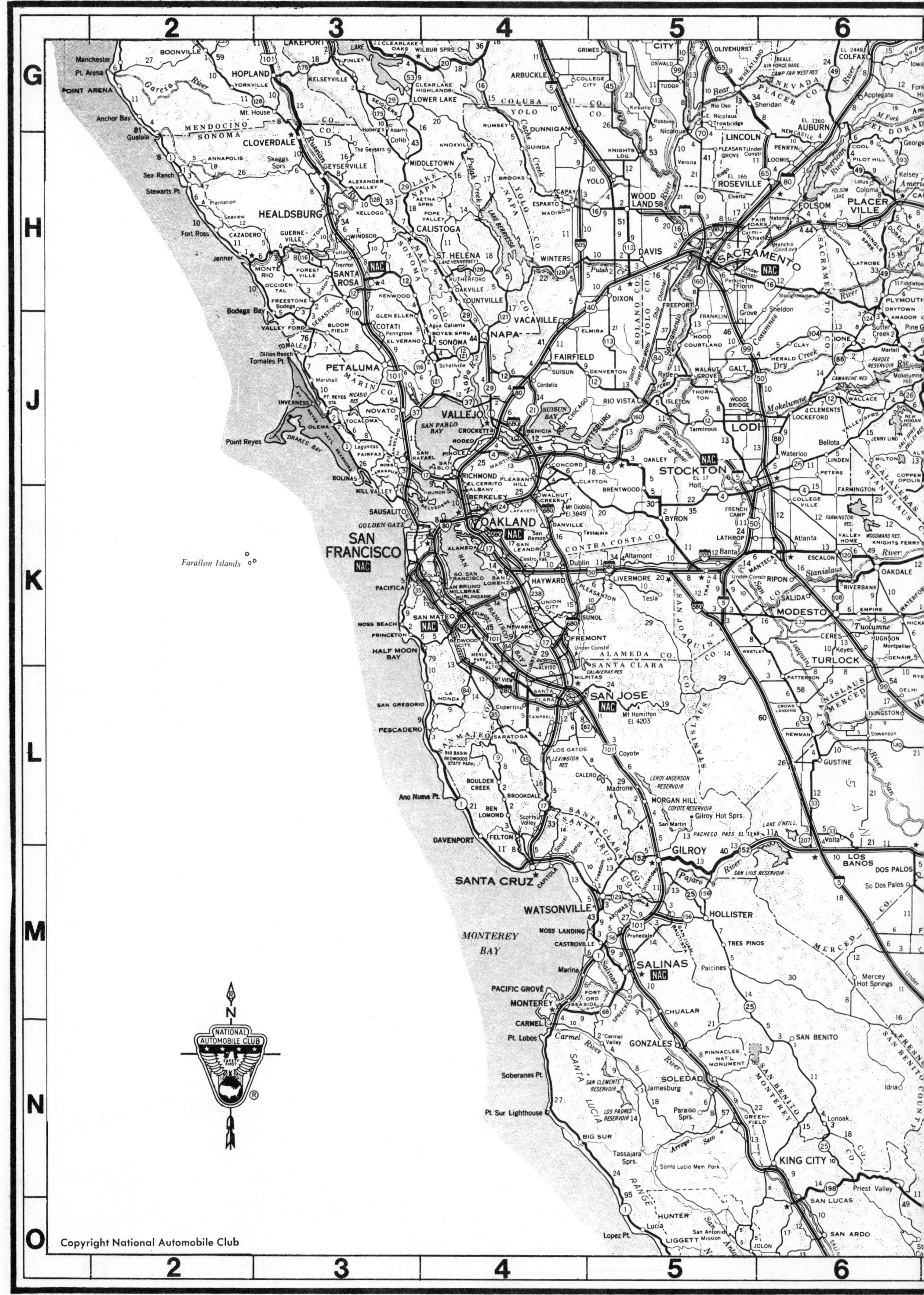

2
3
4
5
6
G
H
J
K
L
M
N
O
POINT ARENA
CLOVERDALE
HEALDSBURG
SANTA ROSA
PETALUMA
NAPA
VALLEJO
VACAVILLE
FAIRFIELD
DAVIS
SACRAMENTO
STOCKTON
LODI
MODESTO
TURLOCK
RICHMOND
BERKELEY
OAKLAND
SAN FRANCISCO
Farallon Islands
Point Reyes
HAYWARD
FREMONT
SAN JOSE
SANTA CRUZ
WATSONVILLE
GILROY
HOLLISTER
SALINAS
MONTEREY
MONTEREY BAY
GONZALES
SOLEDAD
KING CITY
PACIFIC GROVE
CARMEL
BIG SUR
AUBURN
PLACER VILLE
LOS BANOS
NATIONAL AUTOMOBILE CLUB
SAFETY FIRST

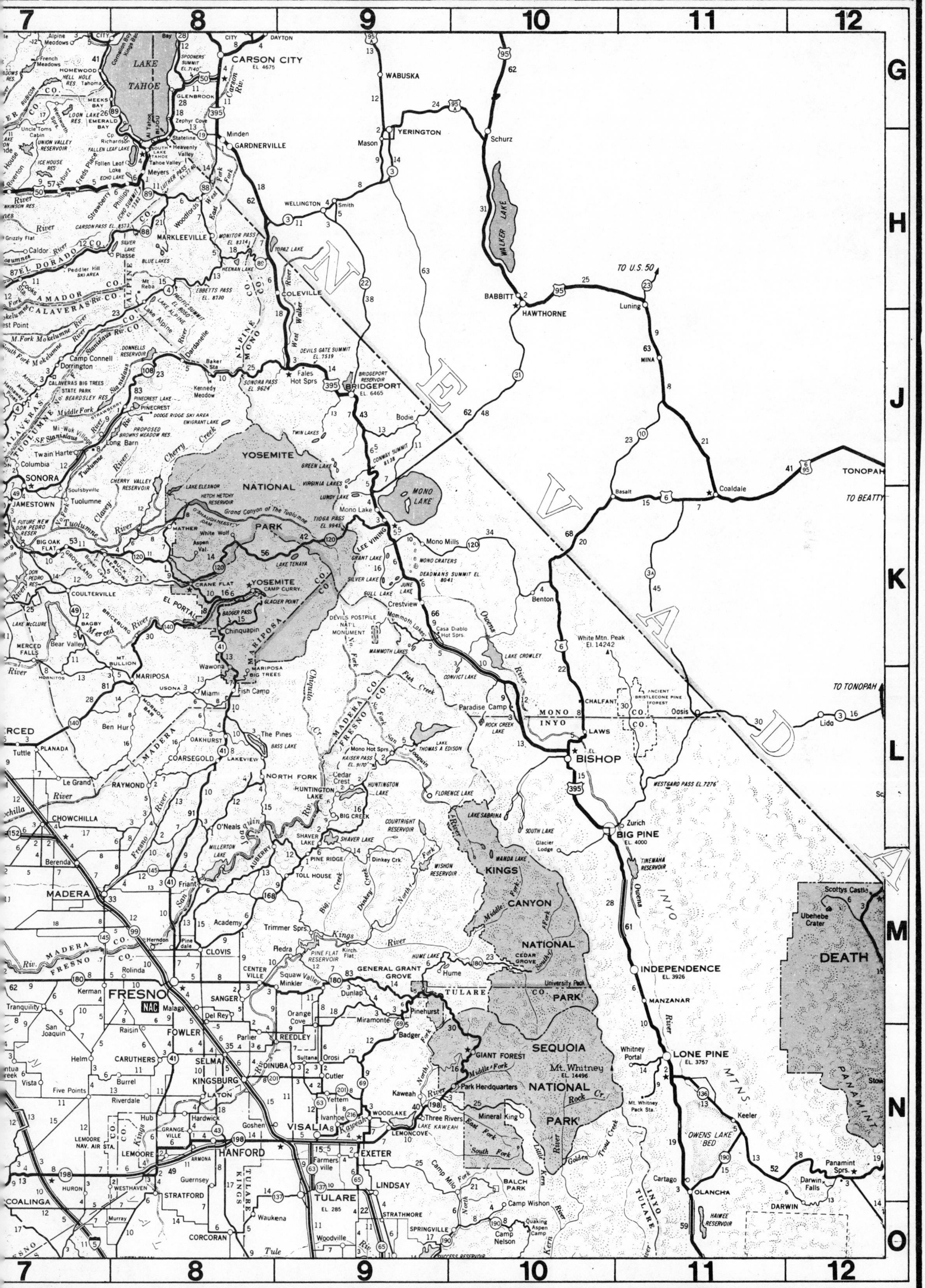

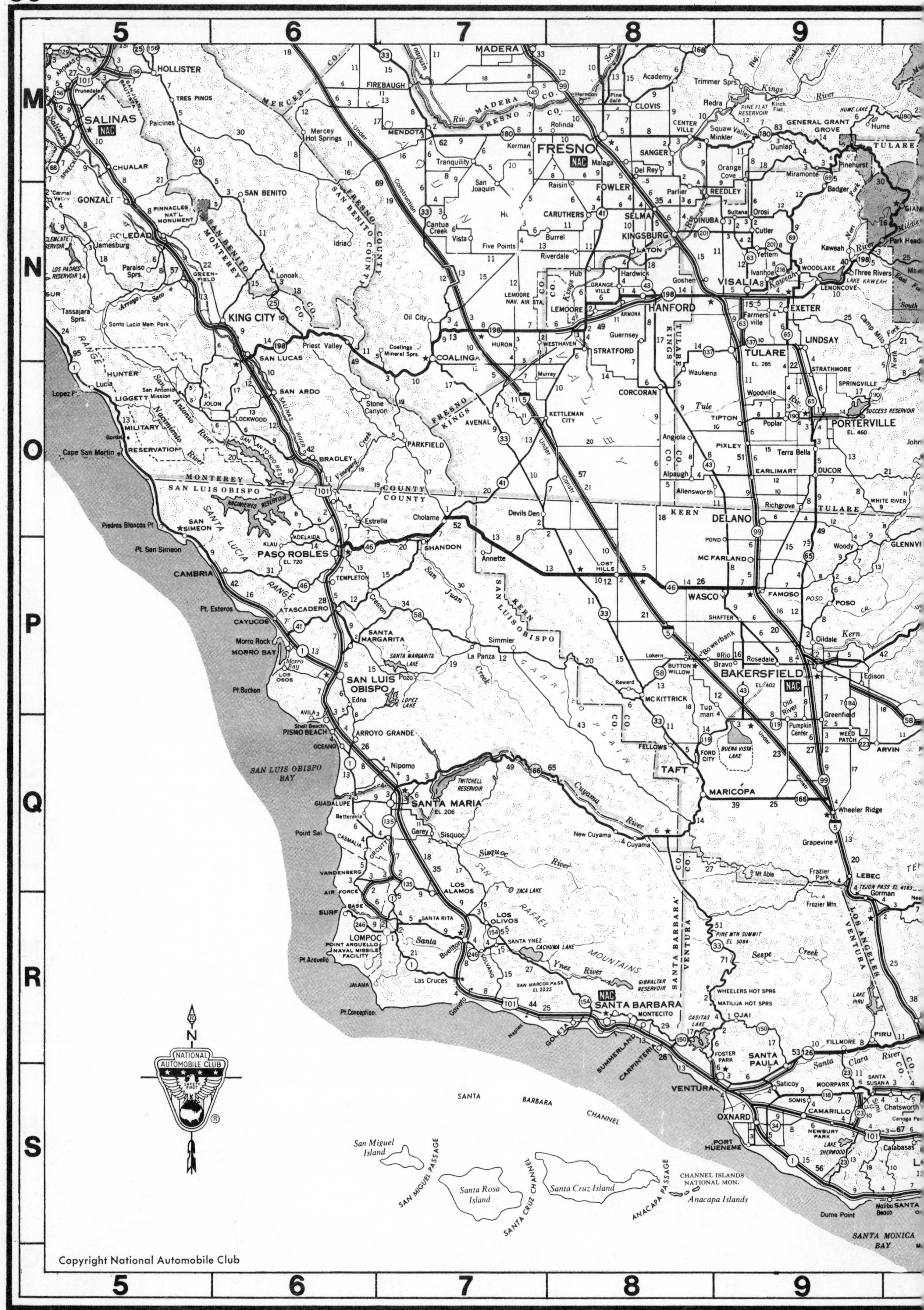
5
6
7
8
9
M
N
O
P
Q
R
S
HOLLISTER
SALINAS
TRES PINOS
Paicines
CHUALAR
GONZALES
SAN BENITO
PINNACLES NAT'L MONUMENT
SOLEDAD
Jamesburg
Paraiso Sprs.
GREENFIELD
KING CITY
Tassajara Sprs.
Lonoak
SAN LUCAS
Priest Valley
HUNTER LIGGETT MILITARY RESERVATION
Lucia
Lopez Pt.
JOLON
LOCKWOOD
SAN ARDO
Cape San Martin
BRADLEY
Stone Canyon
PARKFIELD
COALINGA
Coalinga Mineral Sprs.
Oil City
MONTEREY
SAN LUIS OBISPO
COUNTY
Cholame
SAN SIMEON
Piedras Blancas Pt.
Pt. San Simeon
CAMBRIA
PASO ROBLES
EL. 720
ADELAIDA
KLAU
Estrella
SHANDON
TEMPLETON
Creston
ATASCADERO
Pt. Estero
CAYUCOS
Morro Rock
MORRO BAY
SANTA MARGARITA
SANTA MARGARITA LAKE
SAN LUIS OBISPO
LOS OSOS
Pt.Buchon
Edna
Pozo
LOPEZ LAKE
AVILA
Shell Beach
PISMO BEACH
OCEANO
ARROYO GRANDE
SAN LUIS OBISPO BAY
Nipomo
TWITCHELL RESERVOIR
GUADALUPE
SANTA MARIA
EL. 206
Betteravia
Point Sal
CASMALIA
ORCUTT
Garey
Sisquoc
VANDENBERG AIR FORCE BASE
SURF
LOS ALAMOS
ZACA LAKE
SANTA RITA
LOS OLIVOS
LOMPOC
POINT ARGUELLO NAVAL MISSILE FACILITY
Pt.Arguello
Buellton
SOLVANG
SANTA YNEZ
CACHUMA LAKE
SAN RAFAEL MOUNTAINS
JALAMA
Las Cruces
Pt.Conception
Gaviota
SAN MARCOS PASS EL 2225
GIBRALTAR RESERVOIR
SANTA BARBARA
MONTECITO
GOLETA
SUMMERLAND
CARPINTERIA
VENTURA
SANTA BARBARA CHANNEL
San Miguel Island
SAN MIGUEL PASSAGE
Santa Rosa Island
SANTA CRUZ CHANNEL
Santa Cruz Island
ANACAPA PASSAGE
Anacapa Islands
CHANNEL ISLANDS NATIONAL MON.
MADERA
FIREBAUGH
MENDOTA
Mercey Hot Springs
Kerman
Tranquility
San Joaquin
Cantua Creek
Vista
Five Points
Rolinda
FRESNO
Malaga
CLOVIS
Academy
Herndon
Pinedale
SANGER
Del Rey
FOWLER
Raisin
CARUTHERS
Burrel
Riverdale
SELMA
KINGSBURG
LATON
LEMOORE NAV. AIR STA.
LEMOORE
Hardwick
GRANGEVILLE
ARMONA
HANFORD
HURON
WESTHAVEN
STRATFORD
Guernsey
AVENAL
KETTLEMAN CITY
Murray
CORCORAN
Angiola
Alpaugh
Allensworth
Devils Den
Annette
LOST HILLS
Simmler
La Panza
CUYAMA
New Cuyama
Trimmer Sprs.
Piedra
PINE FLAT RESERVOIR
Kirch Flat
CENTER VILLE
Squaw Valley
Minkler
Orange Cove
Parlier
REEDLEY
DINUBA
Sultana
Orosi
Cutler
Yettem
Ivanhoe
VISALIA
Goshen
Farmersville
EXETER
TULARE
EL. 285
Waukena
LINDSAY
STRATHMORE
Woodville
SPRINGVILLE
SUCCESS RESERVOIR
PORTERVILLE
EL. 460
TIPTON
Poplar
PIXLEY
Terra Bella
EARLIMART
DUCOR
Richgrove
DELANO
WHITE RIVER
Woody
POND
MC FARLAND
WASCO
FAMOSO
SHAFTER
POSO
Oildale
Lokern
BUTTON WILLOW
Bowerbank
Rio Bravo
Rosedale
BAKERSFIELD
EL. 402
Edison
Reward
MC KITTRICK
Tupman
Old River
Greenfield
Pumpkin Center
WEED PATCH
ARVIN
FELLOWS
FORD CITY
TAFT
BUENA VISTA LAKE
MARICOPA
Wheeler Ridge
Grapevine
Mt Able
Frazier Park
Frazier Mtn.
LEBEC
TEJON PASS EL. 4183
Gorman
PINE MTN SUMMIT EL 5084
Sespe Creek
WHEELERS HOT SPRS
MATILIJA HOT SPRS
CASITAS LAKE
OJAI
FOSTER PARK
SANTA PAULA
FILLMORE
PIRU
LAKE PIRU
Saticoy
MOORPARK
SANTA SUSANA
SOMIS
CAMARILLO
Chatsworth
OXNARD
PORT HUENEME
NEWBURY PARK
LAKE SHERWOOD
Calabasas
Dume Point
Malibu Beach
SANTA MONICA BAY
DUNLAP
Miramonte
Pinehurst
Badger
HUME LAKE
Hume
GENERAL GRANT GROVE
Kaweah
WOODLAKE
LEMONCOVE
Three Rivers
LAKE KAWEAH
Camp Nelson
NATIONAL AUTOMOBILE CLUB
SAFETY FIRST

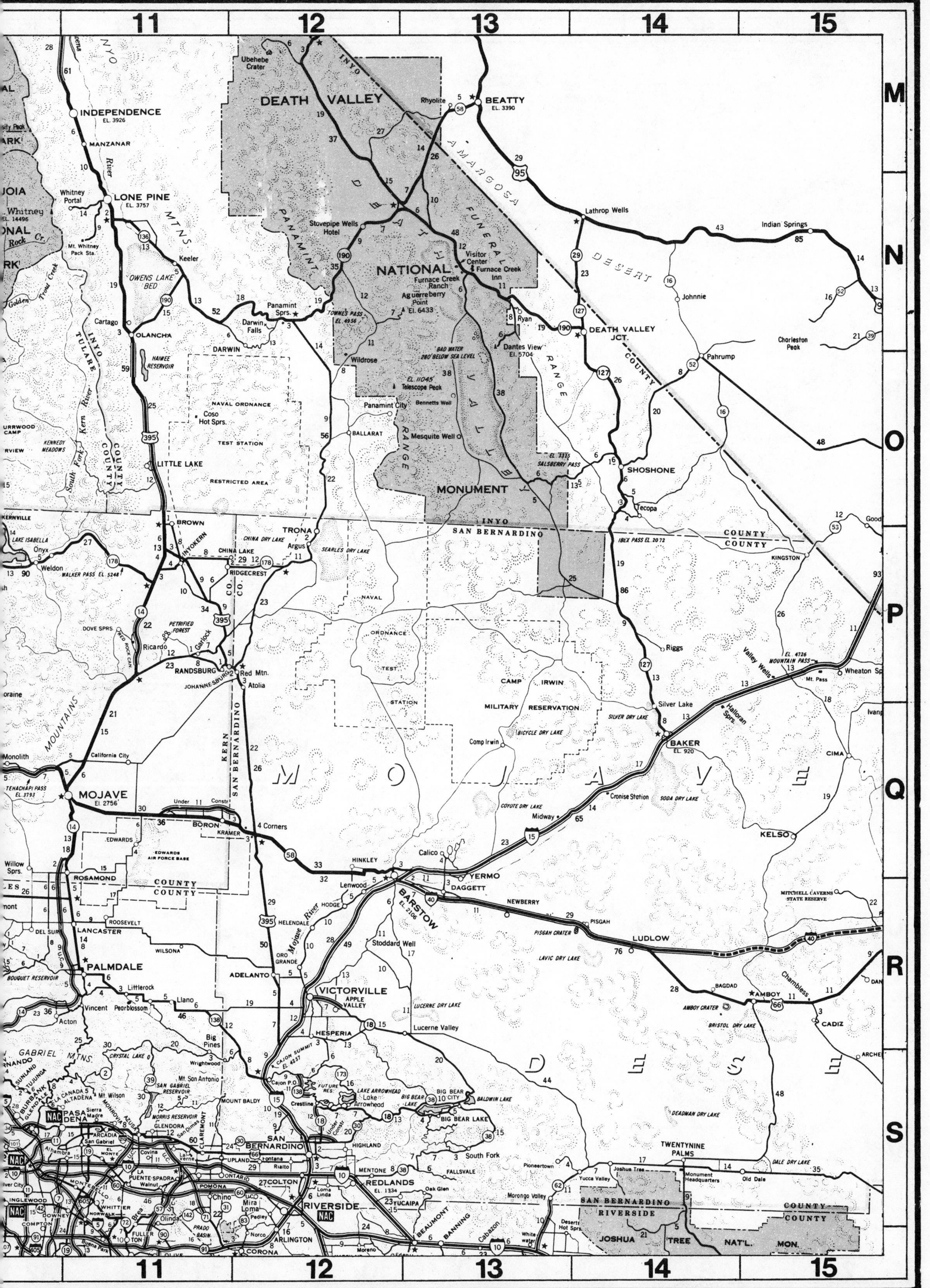

11
12
13
14
15
M
N
O
P
Q
R
S
INYO
Ubehebe Crater
DEATH VALLEY
NATIONAL
MONUMENT
Rhyolite
BEATTY
EL. 3390
INDEPENDENCE
EL. 3926
MANZANAR
River
AMARGOSA
Whitney Portal
LONE PINE
EL. 3757
Mt. Whitney Pack Sta.
Keeler
OWENS LAKE BED
MTNS.
PANAMINT
FUNERAL
Stovepipe Wells Hotel
Visitor Center
Furnace Creek Inn
Furnace Creek Ranch
Aguerreberry Point
EL. 6433
TOWNES PASS EL. 4956
Lathrop Wells
Indian Springs
DESERT
Johnnie
Charleston Peak
Pahrump
Cartago
OLANCHA
Panamint Sprs.
Darwin Falls
DARWIN
HAIWEE RESERVOIR
Ryan
DEATH VALLEY JCT.
Dantes View
EL. 5704
Wildrose
BAD WATER
280' BELOW SEA LEVEL
EL. 11045
Telescope Peak
Bennetts Well
Panamint City
BALLARAT
Mesquite Well
RANGE
COUNTY
NAVAL ORDNANCE
Coso Hot Sprs.
TEST STATION
RESTRICTED AREA
LITTLE LAKE
TULARE
INYO
COUNTY
KENNEDY MEADOWS
South Fork
Kern River
SALSBERRY PASS
EL. 3315
SHOSHONE
Tecopa
BROWN
INYOKERN
CHINA DRY LAKE
CHINA LAKE
RIDGECREST
TRONA
Argus
SEARLES DRY LAKE
SAN BERNARDINO
IBEX PASS EL. 2072
COUNTY
COUNTY
KINGSTON
LAKE ISABELLA
Onyx
Weldon
WALKER PASS EL. 5246
DOVE SPRS.
PETRIFIED FOREST
Garlock
Ricardo
RED ROCK CAN.
RANDSBURG
Red Mtn.
JOHANNESBURG
Atolia
NAVAL
ORDNANCE
TEST
STATION
CAMP IRWIN
MILITARY RESERVATION
Riggs
Valley Wells
EL. 4726 MOUNTAIN PASS
Mt. Pass
Wheaton Sp
Silver Lake
SILVER DRY LAKE
Halloran Sprs.
BICYCLE DRY LAKE
Camp Irwin
BAKER
EL. 920
CIMA
Monolith
MOUNTAINS
California City
KERN
SAN BERNARDINO
TEHACHAPI PASS
EL. 3793
MOJAVE
EL. 2756
M O J A V E
Cronise Station
SODA DRY LAKE
COYOTE DRY LAKE
Midway
KELSO
Under Constr
BORON
KRAMER
4 Corners
EDWARDS
EDWARDS AIR FORCE BASE
HINKLEY
Calico
YERMO
DAGGETT
Willow Sprs.
ROSAMOND
COUNTY
COUNTY
Lenwood
BARSTOW
EL. 2106
NEWBERRY
MITCHELL CAVERNS STATE RESERVE
HODGE
PISGAH
PISGAH CRATER
LUDLOW
LANCASTER
ROOSEVELT
HELENDALE
Mojave River
Stoddard Well
LAVIC DRY LAKE
WILSONA
ORO GRANDE
PALMDALE
BOUQUET RESERVOIR
Littlerock
ADELANTO
VICTORVILLE
APPLE VALLEY
BAGDAD
AMBOY
AMBOY CRATER
Chambless
CADIZ
Vincent
Pearblossom
Llano
LUCERNE DRY LAKE
Lucerne Valley
BRISTOL DRY LAKE
Acton
HESPERIA
GABRIEL MTNS.
CRYSTAL LAKE
Big Pines
Wrightwood
CAJON SUMMIT EL. 4237
D E S E
Mt. San Antonio
SAN GABRIEL RESERVOIR
Mt. Wilson
Cajon P.O.
FUTURE RES.
LAKE ARROWHEAD
Lake Arrowhead
BIG BEAR LAKE
BIG BEAR CITY
BALDWIN LAKE
MOUNT BALDY
Crestline
Big Bear Lake
DEADMAN DRY LAKE
BURBANK
GLENDALE
PASADENA
Sierra Madre
MONROVIA
AZUSA
MORRIS RESERVOIR
GLENDORA
San Dimas
CLAREMONT
SAN BERNARDINO
HIGHLAND
Alhambra
San Gabriel
ARCADIA
EL MONTE
Covina
La Verne
UPLAND
Fontana
Rialto
South Fork
TWENTYNINE PALMS
DALE DRY LAKE
Pioneertown
Yucca Valley
Joshua Tree
Monument Headquarters
Old Dale
PUENTE-SPADRA
Walnut
ONTARIO
COLTON
MENTONE
REDLANDS
EL. 1334
FALLSVALE
INGLEWOOD
WHITTIER
POMONA
Chino
Loma Linda
YUCAIPA
Oak Glen
Morongo Valley
SAN BERNARDINO
RIVERSIDE
COUNTY
COUNTY
DOWNEY
NORWALK
Mira Loma
Pedley
RIVERSIDE
BEAUMONT
BANNING
Cabazon
Desert Hot Sprs.
COMPTON
Brea
FULLERTON
Olinda
PRADO BASIN
Norco
ARLINGTON
CORONA
Moreno
White water
JOSHUA TREE NAT'L. MON.
Buena Park
OLIVE
11
12
13
14
15

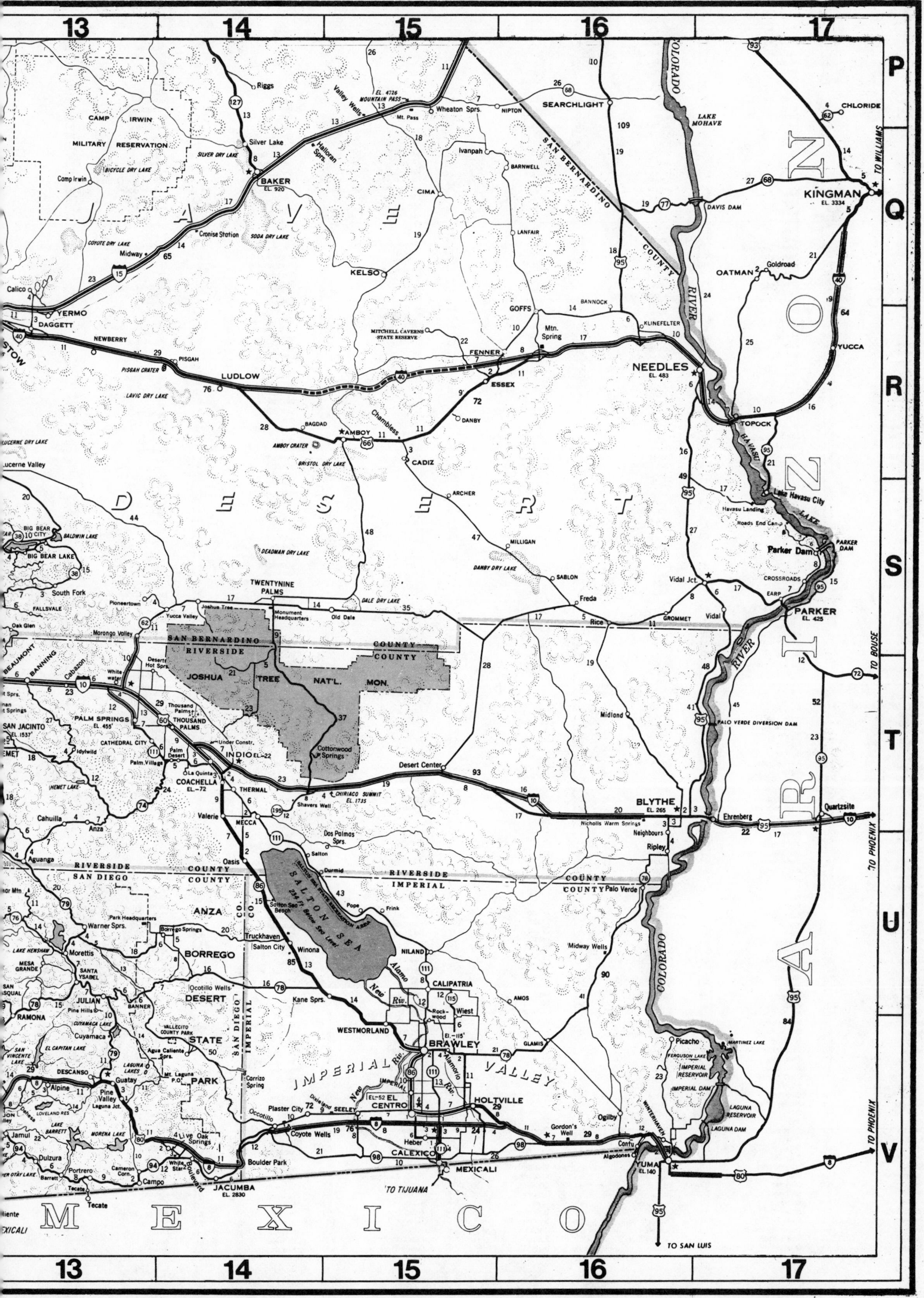

MEXICO
NEVADA
ARIZONA
MOJAVE
DESERT
SALTON SEA
IMPERIAL VALLEY
ANZA DESERT STATE PARK
JOSHUA TREE NAT'L. MON.
KINGMAN
NEEDLES
BLYTHE
PARKER
YUMA
BRAWLEY
EL CENTRO
CALEXICO
MEXICALI
INDIO
PALM SPRINGS
BAKER
TO PHOENIX
TO WILLIAMS
TO BOUSE
TO SAN LUIS
TO TIJUANA

L

M

N

O

P

Q

R

S

T

U

V

W

Index

Y

Z

APPENDIX: County Data

COUNTY	DATE OF ORIGIN	COUNTY SEAT	AREA	1900	1910	1920	1930	1940	1950	1960
Alameda	1853	Oakland	469,120	130,197	246,131	344,177	474,883	513,011	740,315	893,560
Alpine	1864	Markleeville	462,720	509	309	243	241	323	241	370
Amador	1854	Jackson	380,160	11,116	9,086	7,793	8,494	8,973	9,151	9,970
Butte	1850	Oroville	1,065,600	17,417	27,301	30,030	34,093	42,840	64,930	81,940
Calaveras	1850	San Andreas	657,920	14,200	9,171	6,183	6,008	8,221	9,902	10,190
Colusa	1850	Colusa	737,920	7,364	7,732	9,290	10,258	9,788	11,651	12,090
Contra Costa	1850	Martinez	469,760	18,046	31,674	53,889	78,608	100,450	298,984	404,100
Del Norte	1857	Crescent City	641,920	2,408	2,417	2,759	4,739	4,745	8,078	17,480
El Dorado	1850	Placerville	1,104,000	8,986	7,492	6,426	8,325	13,229	16,207	29,280
Fresno	1856	Fresno	3,830,400	37,862	76,657	128,779	144,379	178,565	276,515	368,490
Glenn	1891	Willows	842,880	5,150	7,172	11,853	10,935	12,195	15,448	17,210
Humboldt	1853	Eureka	2,286,720	27,104	33,857	37,413	43,233	45,812	69,241	105,170
Imperial	1907	El Centro	2,741,760	--	13,591	43,453	60,903	59,740	62,975	71,030
Inyo	1866	Independence	6,458,240	4,377	6,974	7,031	6,555	7,625	11,658	11,710
Kern	1866	Bakersfield	5,228,800	16,480	37,715	54,843	82,570	135,124	228,309	294,760
Kings	1893	Hanford	892,000	9,871	16,230	22,031	25,385	35,168	46,768	49,900
Lake	1861	Lakeport	803,840	6,017	5,526	5,402	7,166	8,069	11,481	13,680
Lassen	1864	Susanville	2,910,720	4,511	4,802	8,507	12,589	14,479	18,474	13,560
Los Angeles	1850	Los Angeles	2,605,440	170,298	504,131	936,455	2,208,492	2,785,643	4,151,687	6,011,140
Madera	1893	Madera	1,374,720	6,364	8,368	12,203	17,164	23,314	36,964	40,420
Marin	1850	San Rafael	333,440	15,702	25,114	27,342	41,648	52,907	85,619	146,050
Mariposa	1850	Mariposa	931,200	4,720	3,956	2,775	3,233	5,605	5,145	5,090
Mendocino	1850	Ukiah	2,246,400	20,465	23,929	24,116	23,505	27,864	40,854	50,850
Merced	1855	Merced	1,269,120	9,215	15,148	24,579	36,748	46,988	69,780	90,380
Modoc	1874	Alturas	2,620,160	5,076	6,191	5,425	8,038	8,713	9,678	8,360
Mono	1861	Bridgeport	1,948,800	2,167	2,042	960	1,360	2,299	2,115	2,160
Monterey	1850	Salinas	2,127,360	12,380	24,146	27,980	53,705	73,032	130,498	197,870
Napa	1850	Napa	505,600	16,451	19,800	20,678	22,897	28,503	46,603	65,620
Nevada	1851	Nevada City	626,560	17,789	14,955	10,850	10,596	19,283	19,888	21,050
Orange	1889	Santa Ana	500,480	19,696	34,436	61,375	118,674	130,760	216,224	708,940
Placer	1851	Auburn	915,840	15,786	18,237	18,584	24,468	28,108	41,649	56,960
Plumas	1854	Quincy	1,644,800	4,657	5,259	5,681	7,913	11,548	13,519	11,630
Riverside	1893	Riverside	4,594,560	17,897	34,696	50,297	81,024	105,524	170,046	303,360
Sacramento	1850	Sacramento	630,400	45,915	67,806	91,029	141,999	170,333	277,140	502,770
San Benito	1874	Hollister	893,440	6,633	8,041	8,995	11,311	11,392	14,370	15,300
San Bernardino	1853	San Bernardino	12,883,840	27,929	56,706	73,401	133,900	161,108	281,642	501,130
San Diego	1850	San Diego	2,725,120	35,000	61,665	112,248	209,659	289,348	556,808	1,033,380
San Francisco	1850	San Francisco	28,800	342,782	416,912	506,676	634,394	634,536	775,357	729,180
San Joaquin	1850	Stockton	902,400	35,452	50,731	79,905	102,940	134,207	200,750	248,850
San Luis Obispo	1850	San Luis Obispo	2,128,640	16,637	19,383	21,893	29,613	33,246	51,417	80,670
San Mateo	1856	Redwood City	290,560	12,094	26,585	36,781	77,405	111,782	235,659	441,490
Santa Barbara	1850	Santa Barbara	1,756,800	18,934	27,738	41,097	65,167	70,555	98,220	168,520
Santa Clara	1850	San Jose	835,200	60,216	83,539	100,676	145,118	174,949	290,547	642,160
Santa Cruz	1850	Santa Cruz	280,960	21,512	26,140	26,269	37,453	45,057	66,534	82,410
Shasta	1850	Redding	2,461,440	17,318	18,920	13,361	13,927	28,800	36,413	59,290
Sierra	1852	Downieville	613,120	4,017	4,098	1,783	2,422	3,025	2,410	1,980
Siskiyou	1852	Yreka	4,040,320	16,962	18,801	18,545	25,480	28,598	30,733	32,980
Solano	1850	Fairfield	529,280	24,143	27,559	40,602	40,834	49,118	104,833	134,030
Sonoma	1850	Santa Rosa	1,010,560	38,480	48,394	52,090	62,222	69,052	103,405	146,550
Stanislaus	1854	Modesto	963,840	9,550	22,522	43,557	56,641	74,866	127,231	156,700
Sutter	1850	Yuba City	388,480	5,886	6,328	10,115	14,618	18,680	26,239	33,350
Tehama	1856	Red Bluff	1,903,360	10,996	11,401	12,882	13,866	14,316	19,276	25,450
Trinity	1850	Weaverville	2,042,240	4,383	3,301	2,551	2,809	3,970	5,087	10,040
Tulare	1852	Visalia	3,100,800	18,375	35,440	59,031	77,442	107,152	149,264	166,150
Tuolumne	1850	Sonora	1,456,000	11,166	9,979	7,768	9,271	10,887	12,584	14,060
Ventura	1873	Ventura	1,188,480	14,967	18,347	28,724	54,976	69,685	114,647	199,270
Yolo	1850	Woodland	661,760	13,618	13,296	17,105	23,644	27,243	40,640	66,040
Yuba	1850	Marysville	408,320	8,620	10,042	10,375	11,331	17,034	24,420	33,880
STATE				1,485,053	2,377,549	3,426,861	5,677,883	6,907,387	10,586,223	15,650,000

Table of Distances

CITY	SAN FRANCISCO	SACRAMENTO	LOS ANGELES	SAN DIEGO
Alturas	376	312	661	751
Auburn	121	34	419	541
Bakersfield	293	113	235	272
Banning	516	468	83	130
Barstow	432	407	134	184
Bishop	316	285	269	361
Blythe	624	608	223	243
Brawley	630	574	197	132
Bridgeport	278	191	360	452
Capistrano Beach	493	446	60	61
Carmel	128	198	330	452
Chico	168	100	489	613
Corning	175	115	501	623
Crescent City	372	387	805	927
Death Valley National	483	452	311	361
El Centro	643	596	210	118
Escondido	542	506	109	31
Eureka	285		719	841
Fresno	186	166	219	341
Garberville	213	264	646	768
Gorman	336	315	70	192
Indio	560	513	127	128
King City	151	230	282	404
Klamath Falls, Oregon	367	307	693	815
Lake Tahoe	195	108	445	537
Lassen National Park	244	176	565	689
Lee Vining	248	227	335	427
Lone Pine	375	334	212	304
Manteca	77	56	329	451
Marysville	123	53	442	566
Medford, Oregon	372	313	698	820
Merced	132	111	274	396
Modesto	93	73	312	434
Mojave	354	333	99	221
Monterey	124	194	333	455
Oakland	10	85	395	517
Oceanside, via 101	519	471	86	36
Palm Springs	508	492	107	140
Paso Robles	204	276	229	351
Petaluma	40	89	473	595
Placerville	129	42	429	551
Pomona	462	414	29	129
Portland, Oregon	668	608	994	116
Red Bluff	193	134	519	641
Redding	224	164	550	672
Reno, Nevada	224	137	470	562
Riverside	455	439	54	106
Sacramento	87	--	385	507
Salinas	105	184	328	440
San Bernardino	453	437	57	116
San Diego	529	507	124	--
San Fernando	383			363
San Francisco	--	93	433	555
San Jose	48	126	385	507
San Luis Obispo	232	303	200	322
Santa Ana	468	421	35	87
Santa Barbara	338	408	95	217
Santa Cruz	79	159	356	480
Santa Maria	263	334	169	291
Santa Rosa	56	99	489	611
Seattle, Washington	856	796	1182	1304
Sequoia National Park	273	257	223	345
Stockton	82	45	340	462
Tulare	230	176	298	209
Ukiah	117	159	550	672
Ventura	364	395	68	190
Watsonville	97	175	361	483
Weed	294	235	620	742
Woodland	83	24	409	531
Yosemite National Park	193	172	281	403

MOUNTAIN	COUNTY	LOCATION	ELEV
Agua Tibia Mts.	San Diego	33.20 116.55	6140
Amargosa Rg.	Inyo	36.20 116.45	6384
Argus Rg.	Inyo	35.55 117.30	6562
Avawatz Mts.	San Bernardino	35.30 116.20	6156
Big Maria Mts.	Riverside	33.50 114.40	3375
Caliente Rg.	San Luis Obispo	35.05 119.50	5104
Cargo Muchacho Mts.	Imperial	33.03 114.50	2225
Casmalia Hills	Santa Barbara	34.53 120.33	1640
Castle Mts.	San Bernardino	35.20 115.05	5550
Chemehuevi Mts.	San Bernardino	34.37 114.32	3697
Chocolate Mts.	Imperial	33.15 115.30	2090
Cholame Hills	Monterey	35.52 120.30	2880
Chuckwalla Mts.	Riverside	33.38 115.20	4505
Coso Rg.	Inyo	36.10 117.45	8160
Coxcomb Mts.	San Bernardino	34.00 115.25	4450
Diablo Rg.	Santa Clara	36.45 121.10	5258
El Paso Mts.	San Bernardino	35.25 117.50	5259
Elsinore	Riverside	33.37 117.23	3591
Funeral Mts.	Inyo	36.30 116.40	6703
Gabilan Rg.	San Benito	36.35 121.15	3262
Grapevine Mts.	Inyo	36.55 117.10	8739
Greenhorn Mts.	Tulare (Kern)	35.43 118.35	8320
Grizzly Mts.	Plumas	40.00 120.45	7878
Inyo Mts.	Inyo	36.50 118.00	11123
Iron Mts.	San Bernardino	34.10 115.12	3350
Ivanpah Mts.	San Bernardino	35.23 115.30	6152
Kettleman Hills	Kings	36.00 120.00	1326
Kiavah Mts.	Kern	35.30 118.08	7290
Kingston Rg.	San Bernardino	35.45 115.55	7320
La Panza Rg.	San Luis Obispo	35.15 120.10	4054
Lassen Peak	Shasta	40.30 121.32	10437
Last Chance Rg.	Inyo	37.00 117.35	8726
Little Maria Mts.	Riverside	33.55 114.55	3035
Little San Bernardino Mts.	Riverside	34.00 116.15	5461
Marble Mts.	Siskiyou	41.40 123.12	8317
Mono Craters	Mono	37.51 119.00	9169
Mt. Diablo	Contra Costa	37.53 121.55	3849
Mt. Shasta	Siskiyou	41.25 122.12	14161
Mt. Whitney	Tulare	36.34 118.18	14495
New York Mts.	San Bernardino	35.15 115.18	7445
Nopah Rg.	Inyo	36.00 116.05	6415
Old Woman Mts.	San Bernardino	34.30 115.12	5350
Orocopia Mts.	Riverside	33.35 115.45	3815
Owlshead Mts.	San Bernardino	35.47 116.45	4408
Palen Mts.	Riverside	33.50 115.03	3851
Panamint Rg.	Inyo	36.20 117.08	11049
Pinto Mts.	Riverside	34.00 115.50	4650
Piute Mts.	Kern	35.28 118.23	8433
Providence Mts.	San Bernardino	34.55 115.33	6650
Purisima Hills	Santa Barbara	34.43 120.18	1984
Quail Mts.	San Bernardino	35.37 116.53	5103
Rand Mts.	San Bernardino	35.20 117.40	4755
Red Hills	San Luis Obispo	35.36 120.16	2592
Reef Ridge	Kings	35.55 120.10	4349
Riverside Mts.	Riverside	34.03 114.30	2252
Salmon Mts.	Siskiyou	41.15 123.23	6934
San Bernardino Mts.	San Bernardino	34.10 116.40	11502
San Gabriel Mts.	Los Angeles	34.24 118.00	10059
San Jacinto Mts.	Riverside	33.45 116.38	10831
San Rafael Mts.	Santa Barbara	34.34 119.45	6880
San Ysidro Mts.	San Diego	32.35 116.50	3572
Santa Ana Mts.	Riverside	33.45 117.30	5696
Santa Cruz Mts.	Santa Cruz	37.05 121.50	3798
Santa Lucia Rg.	Monterey	36.15 121.45	5844
Santa Margarita Mts.	Orange	33.27 117.25	3189
Santa Monica Mts.	Los Angeles	34.05 118.50	2836
Santa Rosa Mts.	Riverside	33.33 116.20	8716
Santa Susana Mts.	Los Angeles	34.20 118.35	3756
Santa Ynez Mts.	Santa Barbara	34.30 120.00	4864
Scott Bar Mts.	Siskiyou	41.15 122.55	6265
Scott Mts.	Siskiyou	41.20 122.40	8551
Sheep Hole Mts.	San Bernardino	34.12 115.40	4600
Sierra de Salinas	Monterey	36.25 121.28	4465
Sierra Madre Mts.	Santa Barbara	34.55 119.50	5845
Simi Hills	Ventura	34.17 118.43	2403
Siskiyou Mts.	Siskiyou	41.55 123.25	7310
Slate Rg.	San Bernardino	35.45 117.17	5584
Sutter (Marysville) Buttes	Sutter	39.13 121.49	2132
Tehachapi Mts.	Kern	34.55 118.40	7988
Telescope Peak	Inyo	36.10 117.05	11049
Temblor Rg.	Kern	35.15 119.45	4332
Topatopa Mts.	Ventura	34.31 119.05	6704
Trinity Alps	Trinity	41.00 123.00	8964
Trinity Mts.	Trinity	40.55 122.38	7241
Turtle Mts.	San Bernardino	34.18 114.50	4313
Vallecito Mts.	San Diego	33.03 116.15	5348
Warner Rg.	Modoc	41.25 120.15	9722
Whipple Mts.	San Bernardino	34.19 114.25	4131
White Mts.	Mono - Inyo	37.35 119.15	14246
Yolla Bolly Mts.	Tehama	40.02 122.50	8083

Passes

PASS OR SUMMIT	COUNTY	ROUTE NO.	ELEV
Adin Summit	Modoc	US 299	5173
Avawatz Pass	San Bernardino	--	4169
Bad Water (Death Valley)	Inyo	--	280
Beckworth Pass	Plumas	US 40 Alt.	5212
Berry Summit	Humboldt	US 299	2871
Black Butte	Siskiyou	US 99	3920
Blue Ridge Summit	Los Angeles	SSR 2	7378
Buckhorn Mt. Summit	Trinity	US 299	3213
Cadiz Summit	San Bernardino	US 66	1275
Cajon Pass	San Bernardino	US 66, 91, 395	4301
Carson Pass	Alpine	SSR 88	8573
Casitas Pass, East	Ventura	SSR 150	1158
Cedar Pass	Modoc	US 299	6350
Cherry Gap	Fresno	SSR 180	6897
Cloudburst Summit	Los Angeles	SSR 2	7018
Conway Summit	Mono	US 395	8138
Cottonwood Pass	Madera	SSR 41	2000
Cushenbury Grade	San Bernardino	SSR 18	6000
Dawson Saddle	Los Angeles	SSR 2	7901
Daylight Pass	Inyo	Nev. 58	4315
Dead Horse Summit	Siskiyou	SSR 89	4535
Dead Mans Summit	Mono	US 395	8040
Deer Mt.	Siskiyou	US 97	5202
Deer Creek Pass	Tehama	SSR 36	4939
Devils Gate Pass	Mono	US 395	7519
Donner Summit	Nevada	US 40	7135
Ebbetts Pass	Alpine	SSR 4	8730
Echo Summit	El Dorado	US 50	7382
Fandango Pass	Modoc	--	6100
Fredonyer Pass	Lassen	SSR 36	5748
Gaviota Gorge	Santa Barbara	US 101	200
Gaviota Pass	Santa Barbara	US 101	200
Gilbert Pass	Inyo	--	6371
Hatchet Mt. Summit	Shasta	US 299	4368
Hayfork Summit	Trinity	US 299	3660
Hazelview Summit	Del Norte	US 199	2435
Hecker Pass	Santa Cruz	SSR 152	1309
Holland Summit	Los Angeles	US 99	4000
Ibex Pass	San Bernardino	SSR 127	2250
Islip Saddle	Los Angeles	SSR 2	6658
Jarboe Pass	Butte	US 40 Alt.	2330
Jubilee Pass	Inyo	SSR 212	1545
Kaiser Pass	Fresno	--	9250
Kratka Ridge	Los Angeles	SSR 2	
La Cuesta Pass	San Luis Obispo	US 101	1550
Little Truckee Summit	Sierra	SSR 89	1025
Luther Pass	El Dorado	SSR 89	7740
McCoy Saddle	Tuolumne	SSR 108	5580
Mineral Summit	Tehama	SSR 36	5265
Monitor Pass	Alpine-Mono	SSR 89	8314
Mono Pass	Inyo	--	12050
Montgomery Pass	--	US 6	7122
Morgan Summit	Tehama	SSR 36	5750
Mt. Hebren Summit	Siskiyou	US 97	5202
Mt. Pass	San Bernardino	US 91, 466	4731
Mtn. Pass	Tuolumne	SSR 108	1405
Nojoqui Pass	Santa Barbara	US 101	925
O'Brien Summit	Shasta	US 99	1000
Oregon Mt.	Trinity	US 299	2045
Pacheco Pass	Santa Clara	SSR 152	1386
Pacific Grade Summit	Alpine	SSR 4	8050
Panoche Pass	San Benito	--	2150
Peddler Hill Summit	Amador	SSR 88	7000
Pine Mtn. Summit	Ventura	US 399	5080
Pine Ridge	Fresno	SSR 168	4500
Rattlesnake Summit	Mendocino	US 101	2200
Ridgewood Summit	Mendocino	US 101	1956
Salsbury Pass	Inyo	SSR 212	3315
Salton Sea	Imperial	US 99	-244
San Gorgonio Pass	San Bernardino	US 60, 70	2612
San Marcos Pass	San Bernardino	SSR 150	2225
Shavers Summit	Riverside	US 60, 70	1740
Sherwin Summit	Mono	US 395	7000
Sherwin Hill Summit	Mono	US 395	6000
Simmler Summit	Kern	SSR 178	2047
Sonora Pass	Mono	SSR 108	9626
South Pass	Santa Barbara	US 66	2750
Southfork Mt.	Trinity	SSR 36	4073
Tamarack Ridge	Fresno	SSR 168	7582
Tecate Divide	San Diego	US 80	3890
Tehachapi Summit	Kern	US 466	3988
Tejon Pass	Los Angeles	US 99	4183
The Lord Ellis Summit	Humboldt	US 299	590
Tioga Pass	Tul	SSR 140	9941
Townes Pass	Inyo	SSR 190	4956
Walker Pass	Kern	SSR 178	5250
Westgard Pass	Inyo	SSR 63	7271
Wheeler Ridge	Kern	US 99	941
Whitaker Summit	Los Angeles	US 99	2950
Yuba Pass	Sierra	SSR 49	6701

APPENDIX: Climatic Data — Temperature

Station		J	F	M	A	M	J	J	A	S	O	N	D
Alturas	Max.	39	43	51	61	69	76	87	86	79	66	53	63
	Min.	16	19	25	30	36	42	45	41	35	29	23	20
Bakersfield		56	63	71	78	85	92	102	101	93	82	70	58
		35	40	44	49	56	62	69	67	60	51	41	36
Barstow		60	65	71	80	88	95	103	102	96	84	71	62
		30	34	39	47	54	60	66	64	58	48	36	32
Big Bear		42	43	48	56	85	71	77	77	72	62	52	45
		15	17	20	26	33	39	45	45	39	31	23	20
Bishop		54	57	64	72	81	89	97	95	89	77	64	55
		22	26	31	38	44	49	54	52	46	38	28	24
Blythe		67	72	80	88	96	103	109	107	103	91	77	69
		36	41	46	53	59	66	75	75	68	55	42	38
Bridgeport		38	41	48	59	66	74	84	83	77	65	53	44
		9	12	19	27	33	38	45	44	37	29	21	15
Burbank		65	66	69	73	76	80	87	88	86	79	74	67
		40	42	45	48	52	55	59	59	57	51	45	42
Chico		53	59	65	73	81	89	97	95	90	78	65	54
		35	38	41	45	51	56	61	58	55	48	40	37
Coalinga		57	62	67	75	83	89	98	96	91	80	67	58
		34	38	41	46	52	57	64	61	57	49	40	37
Crescent City		52	54	55	59	61	64	66	66	66	63	60	55
		37	40	40	41	45	48	50	50	49	46	43	41
Davis		54	59	65	73	81	88	95	94	90	79	66	55
		35	39	41	44	49	53	55	53	52	47	40	37
Escondido		66	67	70	73	76	80	88	88	87	80	75	68
		36	39	42	47	51	54	58	59	56	49	40	38
Eureka		54	55	55	56	58	60	61	61	62	60	58	55
		41	42	43	45	48	51	52	53	51	48	45	42
Fort Bragg		56	57	58	59	62	63	64	64	65	64	61	58
		39	40	41	44	46	49	49	49	49	47	43	42
Fresno		54	61	68	76	84	92	100	98	91	79	66	55
		36	40	43	48	53	59	64	61	56	49	41	38
Giant Forest		43	44	50	55	61	68	77	77	74	62	52	44
		23	23	26	31	36	43	51	50	46	38	31	26
Greenland Ranch		66	72	82	91	100	108	115	113	106	92	77	68
		38	44	52	62	71	79	87	84	75	61	46	39
Indio		70	73	80	87	95	101	107	106	102	92	80	72
		37	42	49	57	65	71	77	76	69	58	45	39
King City		62	64	70	75	80	83	87	85	87	81	73	64
		35	38	39	42	46	48	51	50	48	44	37	36
Long Beach		65	65	67	69	72	74	79	80	79	75	72	67
		45	47	49	52	55	58	62	62	61	56	50	47
Los Angeles		65	66	69	71	74	77	83	84	82	77	73	67
		45	47	49	52	55	58	62	62	60	56	51	48
L. A. Int. Airport		63	63	65	67	69	72	74	75	75	72	70	66
		43	45	48	52	55	58	61	61	59	55	48	45
Markleeville		46	48	56	63	67	76	86	87	81	69	55	46
		14	15	--	26	30	36	40	39	41	26	19	15
Merced		54	60	66	75	83	90	97	95	90	79	66	55
		35	38	40	45	51	56	60	58	54	47	38	36
Modesto		53	59	65	73	80	87	94	91	87	77	65	54
		36	39	42	46	51	55	58	57	54	48	40	38
Needles		61	67	76	86	95	104	109	106	100	87	73	62
		39	43	48	55	62	69	80	78	69	57	45	41
Oakland		56	60	63	66	69	72	72	72	74	71	64	57
		38	41	43	45	48	52	53	54	53	49	43	40
Oroville		54	59	65	73	82	90	98	97	92	79	66	56
		36	39	42	46	52	57	60	58	55	49	41	38
Oxnard		64	65	67	68	69	71	73	74	74	73	71	66
		42	42	44	46	49	52	55	55	54	50	45	43
Palmdale		56	61	64	74	81	88	98	97	92	80	66	57
		28	32	36	43	49	54	63	60	56	55	34	30
Palm Springs		68	71	79	87	94	102	108	106	102	91	79	70
		39	43	47	53	58	63	73	71	66	57	47	41
Paso Robles		61	63	68	75	80	86	93	93	90	82	71	61
		32	35	38	41	44	47	50	49	50	42	35	34
Pt. Piedras		58	57	57	57	58	60	61	61	62	61	62	59
		45	45	46	46	48	49	51	51	51	50	49	47
Porterville		57	63	69	76	85	92	99	97	92	81	69	58
		35	39	42	47	52	57	63	71	57	49	40	37
Red Bluff		53	59	65	71	81	90	99	97	90	78	65	55
		37	41	44	49	55	62	67	64	60	52	44	39
Redding		54	58	64	71	80	88	97	96	90	77	65	55
		37	41	44	49	55	62	67	64	60	52	44	39
Riverside		65	67	71	75	80	86	94	94	92	83	75	67
		37	39	41	46	50	53	57	66	54	48	41	39
Sacramento		52	59	65	71	78	86	92	90	86	76	65	54
		38	42	45	48	52	56	59	58	57	51	44	40
Salinas		61	62	65	67	69	71	71	72	76	74	70	62
		38	41	42	44	48	50	52	53	51	47	42	40
San Bernardino		66	67	71	76	81	87	97	97	94	84	76	68
		36	39	42	46	49	52	58	57	54	48	41	38
Sandberg		46	48	53	60	67	76	85	85	80	67	57	49
		34	34	37	41	47	54	63	62	59	50	42	36
San Diego		64	65	67	68	70	72	76	77	76	73	71	66
		46	48	50	53	57	60	63	64	62	57	51	47
San Fernando		64	65	70	75	79	84	93	93	90	81	74	65
		43	42	44	47	49	51	55	54	53	50	47	44
San Francisco		55	59	61	62	63	65	64	65	68	68	64	57
		45	47	49	49	51	53	53	54	55	54	51	47
San Francisco Airport		56	59	61	63	65	69	69	70	72	69	64	57
		40	43	44	45	48	50	52	52	52	49	45	42
San Jose		57	61	65	69	74	78	81	80	80	74	66	58
		40	43	45	47	50	53	55	55	54	50	45	42
San Luis Obispo		62	62	65	69	70	73	77	77	78	75	71	63
		41	43	45	46	48	51	53	53	52	51	47	44
Santa Ana		67	68	71	74	77	80	85	86	85	80	75	69
		38	40	42	47	51	54	58	58	56	51	43	40
Santa Barbara		65	65	68	70	72	73	77	78	78	76	73	67
		39	42	44	47	50	52	56	56	55	50	44	42
Santa Catalina		58	59	60	64	65	68	77	79	79	72	66	60
		46	46	46	49	51	53	59	61	62	56	53	49
Santa Maria		64	65	67	70	71	72	73	74	75	74	71	65
		37	40	42	44	47	49	52	52	50	47	42	40
Santa Monica		65	65	66	68	70	72	75	76	75	72	69	66
		46	47	48	51	54	57	59	60	59	55	51	48
Santa Rosa		58	62	67	71	75	80	84	84	84	78	68	58
		36	38	40	42	45	48	50	49	48	45	39	38
Scotia		55	57	59	61	64	67	69	70	71	67	61	56
		41	42	43	46	49	52	54	54	52	49	45	43
Stockton		54	60	66	73	80	86	93	91	88	78	66	55
		35	38	40	44	48	53	56	54	52	47	38	37
Susanville		40	45	53	63	71	79	89	88	80	67	52	43
		20	24	29	34	40	46	52	49	44	37	28	24
Tahoe		36	38	43	51	60	68	79	78	70	58	45	39
		17	18	22	27	31	38	43	43	38	32	25	21
Victorville		57	61	66	76	84	91	99	97	91	79	66	59
		25	30	34	40	46	51	57	57	50	41	31	27
Yosemite Nat'l Park		47	52	59	67	73	80	90	90	83	71	58	47
		26	28	31	37	43	47	54	52	47	39	31	28

	ELEV.	ANN.	J	F	M	A	M	J	J	A	S	O	N	D
Alturas	4365	12.97	1.73	1.46	1.40	1.09	1.27	1.03	.26	.21	.42	1.10	1.34	1.66
Bakersfield	489	6.36	1.02	1.12	1.11	.75	.35	.09	.01	.01	.07	.37	.43	1.03
Barstow	2142	4.72	.85	.59	.82	.20	.01	T	.13	.38	.16	.26	.42	.90
Big Bear	6800	37.19	5.50	7.29	7.08	3.02	.30	10	.67	.44	.85	1.71	2.87	7.36
Bishop	4108	5.76	1.42	.89	.73	.29	.30	.09	.10	.11	.17	.27	.40	.99
Blythe	266	4.33	.49	.51	.46	.15	.02	.03	.24	.81	.38	.27	.28	.69
Bridgeport	6420	10.47	1.57	1.78	1.08	.64	.42	.45	.51	.32	.24	.64	.96	1.86
Burbank	699	13.88	2.35	3.06	2.25	1.21	.27	.07	T	.02	.29	.52	.98	2.86
Chico	230	26.60	4.75	4.56	3.55	2.10	1.15	.42	.01	.07	.20	1.56	2.48	5.75
Coalinga	676	6.92	1.06	1.31	1.34	.58	.21	.04	.01	.01	.07	.26	.49	1.54
Crescent City	45	65.36	10.71	9.30	7.81	4.15	2.97	1.76	1.53	.31	1.24	8.09	7.13	10.36
Davis	51	16.58	3.17	3.15	2.37	1.22	.57	.15	T	T	.05	.87	1.41	3.62
Escondido	660	17.59	2.70	3.56	2.70	1.38	.25	.10	.01	.20	.27	.98	1.42	4.02
Eureka	43	39.53	6.94	5.99	5.26	3.07	1.98	.82	.12	.16	.88	2.74	5.09	6.48
Fort Bragg	80	37:85	6.68	5.88	5.16	2.60	1.64	.64	.06	.01	.31	2.89	4.38	7.62
Fresno	294	9.31	1.57	1.66	1.63	.96	.28	.11	.01	T	.05	.66	.75	1.63
Greenland Ranch	-168	1.90	.14	.29	.21	.17	.07	.01	.11	.19	.12	.10	.16	.33
Indio	11	3.91	.50	.47	.28	.11	.02	.01	.14	.41	.52	.26	.35	.84
King City	320	11.17	2.43	2.37	1.88	.92	.14	.09	.02	.01	.05	.26	.61	2.39
Long Beach	34	13.80	2.31	3.40	2.03	.80	.09	.06	T	.05	.23	.45	1.21	3.17
Los Angeles	312	14.54	2.38	3.37	2.36	1.17	.26	.07	T	.02	.27	.50	1.03	3.11
L. A. Int. Airport		12.37	2.01	2.75	1.91	.96	.30	.07	T	.02	.21	.43	1.10	2.61
Markleeville	5546	19.17	3.68	2.39	1.99	1.48	.75	.61	.29	.32	.19	1.39	2.41	3.67
Modesto	91	12.44	2.17	2.18	2.04	1.10	.45	.07	.02	T	.13	.67	1.01	2.60
Needles	913	4.75	.71	.27	.47	.33	.03	.01	.25	.94	.32	.33	.32	.77
Oakland	440	17.47	3.73	2.94	2.36	1.31	.64	.16	T	.03	.10	.82	1.68	3.70
Oroville	272	28.77	5.38	4.92	4.20	2.10	1.26	.39	.01	.01	.22	1.44	3.11	5.73
Oxnard	45	15.41	3.15	3.08	2.43	.94	.11	.06	T	.04	.11	.48	1.11	3.90
Palmdale	2655	6.12	.82	1.01	.98	.28	.06	.02	.11	.02	.07	.23	.78	1.74
Palm Springs	411	7.07	1.22	1.32	.75	.25	.02	.03	.29	.27	.40	.33	.47	1.72
Paso Robles	700	14.36	3.04	2.62	2.21	1.04	.30	.10	.06	.01	.01	.42	1.09	3.46
Pt. PiedrasBlancas	32	19.28	3.80	3.82	3.51	1.28	.32	.04	.03	.01	.03	.91	1.53	4.00
Porterville	393	11.47	2.16	2.02	2.10	1.14	.31	.09	.01	.01	.02	.54	.89	2.18
Red Bluff	350	21.57	3.73	3.53	2.61	1.79	1.06	.46	.02	.05	.33	1.49	2.27	4.23
Redding	577	38.60	7.14	5.81	5.13	3.10	1.75	1.18	.11	.06	.26	2.44	3.75	7.88
Riverside	820	11.96	1.79	2.58	2.00	.91	.18	.04	.01	.20	.11	.60	.87	2.67
Sacramento	25	18.09	3.62	2.96	2.60	1.48	.70	.13	.01	.01	.24	.81	1.83	3.70
Salinas	70	14.75	2.92	2.83	2.32	1.15	.36	.09	.03	.06	.04	.63	1.24	3.08
San Bernardino	1094	18.97	2.89	4.14	2.98	1.67	.32	.11	.04	.19	.20	.98	1.42	4.03
Sandberg	4517	12.09	2.67	2.29	1.59	.87	.26	.03	.02	.08	.18	.55	.83	2.72
San Diego	19	9.90	1.90	1.94	1.52	.70	.29	.05	.04	.09	.10	.44	.89	1.94
San Fernando	950	18.63	3.64	4.09	3.07	1.27	.18	.10	.01	.05	.24	.54	1.34	4.10
San Francisco	52	21.78	4.70	3.64	3.04	1.51	.68	.15	.01	.02	.28	.95	2.36	4.44
San Francisco Apt.	8	17.98	3.88	3.06	2.65	1.19	.47	.12	.01	.02	.11	.75	1.55	4.17
San Jose	95	13.35	2.59	2.56	2.13	1.03	.39	.05	T	.03	.05	.69	1.09	2.74
San Luis Obispo	300	22.44	4.38	4.50	3.43	1.60	.37	.20	.04	.04	.07	.83	1.73	5.25
Santa Ana	133	15.93	2.70	3.42	2.49	1.20	.23	.05	.02	.06	.24	.60	1.37	3.55
Santa Barbara	120	18.17	3.65	3.80	3.16	1.17	.26	.10	.04	.06	.06	.60	1.23	4.04
Santa Catalina	0	12.88	1.98	2.97	1.88	.93	.26	.07	T	.04	.14	.76	.90	2.95
Santa Maria	238	14.49	3.02	2.49	2.55	.99	.21	.20	.04	.03	.13	.72	1.06	3.05
Santa Monica	0	12.87	2.21	2.64	2.49	.74	.06	.01	.01	.01	.21	.31	1.08	3.10
Santa Rosa	167	29.40	5.74	5.08	4.21	2.14	1.00	.35	.02	.01	.09	1.55	3.15	6.06
Scotia	139	48.71	8.53	7.12	6.56	2.93	1.88	.77	.05	.05	.38	3.88	6.80	9.76
Stockton	11	14.62	2.75	2.76	2.28	1.03	.53	.07	.01	T	.11	.71	1.38	2.99
Susanville	4148	14.92	2.53	2.50	1.62	.88	.80	.72	.15	.09	.29	.97	1.71	2.66
Tahoe	6228	31.35	6.12	5.29	4.07	1.96	1.20	.64	.26	.10	.34	2.10	3.56	5.71
Victorville	2750	6.38	1.13	1.10	1.11	.48	.06	.00	.05	.18	.16	.34	.53	1.24
Yosemite N. Park	3985	37.62	6.41	6.90	5.33	3.17	1.37	.50	.24	.07	.38	2.00	3.84	7.41

NAME	COUNTY	WATER SOURCE	CAPACITY AND USE*
Anderson	Santa Clara	Coyote River	75,000 DIIn
Barrett	San Diego	Cottonwood Creek	44,863 DM
Beardsley	Tuolumne	Mid. Fork Stanislaus	97,500 IP
Bear Valley	Orange	Bear Creek	72,400 IR
Big Creek #7	Fresno	San Joaquin	35,000 P
Big Dry Creek	Fresno	Big Dry Creek	16,250 F
Big Sage	Modoc	Rattlesnake Creek	17,000 I
Black Butte	Tehama	Stony Creek	160,000 IR
Boca	Nevada	Little Truckee	41,200 I
Bouquet Canyon	Los Angeles	Bouquet Creek	36,505 M
Bowman	Nevada	Canyon Creek	68,000 DIMMi
Bridgeport	Mono	East Walker River	42,455 I
Briones	Contra Costa	Bear Creek	67,500 DM
Bucks Creek	Plumas	Bucks Creek	103,000 IP
Buena Vista	Kern	Kern	205,000 I
Bullards Bar	Yuba	N. Fork Yuba	31,489 P
Butt Valley	Plumas	Butt Creek	49,768 P
Cachuma	S. Barbara	Santa Ynez River	210,000 IM
Calaveras	Alameda	Calaveras Creek	100,000 DM
Camanche	San Joaquin	Mokelumne	431,500 M
Camp Far West	Placer	Bear River	150,000 I
Casitas	Ventura	Coyote Creek	250,000 IMDR
Chatsworth	Los Angeles	Los Angeles River	9,886 DM
Cherry Valley	Tuolumne	Cherry River	268,000 DIPM
Clear Lake	Modoc	Lost River	527,000 I
Clear Lake	Lake	Cache Creek	420,000 I
Cogswell	Los Angeles	N. Fk. San Gabriel	10,915 DIP
Coyote	Mendocino	E. Fk. Russian River	122,500 IP
Conn Creek Dam	Napa	Conn Creek	30,000 DIM
Copco #1	Siskiyou	Klamath River	77,000 P
Copper Basin	San Bernard.	Copper Basin	22,000 M
Courtright	Fresno	Helms Creek	123,300 P
Coyote	Mendocino	E. Fk. Russian River	122,500 FI
Coyote	S. Clara	Coyote Creek	27,770 I
Crane Valley	Madera	N. Fk. San Joaquin	45,140 P
Cuyamaca	Orange	Boulder Creek	11,600 DI
Dallas Warner	Stanislaus	Tuolumne River	27,000 I
Donnell's	Tuolumne	Mid. Fk. Stanislaus	64,500 IP
Donner Lake	Nevada	Donner Creek	11,000 IP
Don Pedro	Tuolumne	Tuolumne River	289,000 IP
Dorris	Modoc	Stockdill Slough	11,100 IS
East Park	Colusa	Little Stony Creek	51,000 I
El Capitan	San Diego	San Diego River	116,452 M
Englebright	Nevada	Yuba River	70,000 F
Exchequer	Mariposa	Merced River	289,000 IP
Farmington	San Joaquin	Littlejohn Creek	52,000 F
Florence Lake	Fresno	S. Fk. San Joaquin R.	64,406 P
Folsom	Sacramento	American River	1,000,000 F
French Lake	Nevada	Canyon Creek	12,500 DIPMMi
Friant	Fresno	San Joaquin River	520,000 I
Gem Lake	Mono	Rush Creek	17,604 IP
Gibralter	S. Barbara	Santa Ynez River	15,000 DM
Grant Lake	Mono	Rush Creek	47,525 M
Haiwee	Inyo	Rose Valley	58,525 DIM
Hansen	Los Angeles	Big Tujunga River	35,800 F
Henshaw	San Diego	San Luis Rey River	203,581 I
Hillside	Inyo	S. Fk. Bishop Creek	13,368 IP
Hogan	Calaveras	Calaveras River	76,000 M
Huntington Lake	Fresno	Big Creek	88,834 P
Imperial	Imperial	Colorado	85,000 F
Independence	Nevada	Independence Creek	18,500 DP
Isabella	Kern	Kern River	570,000 F
James H. Turner	Alameda	San Antonio Creek	50,500 DM
Keswick	Shasta	Sacramento	23,800 F
Lake Almanor	Plumas	Feather River	1,308,000 IP
Lake Arrowhead	San Bernard.	Little Bear Creek	47,000 DIPR
Lake Camille	Napa	Tulucay Creek	22,000 I
Lake Curry	Napa	Gordon Valley Creek	10,700 M
Lake Eleanor	Tuolumne	Eleanor Creek	27,800 DPM
Lake Fordyce	Nevada	Fordyce Creek	46,662 P
Lake Hemet	Riverside	San Jacinto River	14,000 DI
Lake Hodges	San Diego	San Dieguito River	33,550 DM
Lake Leavitt	Lassen	Susan River	12,100 I
Lake Loveland	San Diego	Sweetwater River	27,700 DIMIn
Lake Spaulding	Nevada	S. Fk. Yuba River	74,448 P
Lake Tahoe	Placer	Truckee River	732,000 IPR
Lake Van Norden	Nevada	S. Fk. Yuba River	74,448 P
Lewiston	Trinity	Trinity River	14,000 IMDPR
Lexington	Santa Clara	Los Gatos Creek	25,000 I
Little Grass Valley	Plumas	S. Fk. Feather River	93,010 DIP
Long Valley	Mono	Owens River	183,465 M
Lower Bear River	Amador	Bear River	48,500 DIP
Lwr. Crystal Spgs.	San Mateo	San Mateo Creek	54,000 DM
Lwr. San Fernando	Los Angeles	San Fernando Creek	20,518 M
Lwr. San Leandro	Alameda	San Leandro Creek	12,600 DM
McCoy Flat	Lassen	Susan River	17,290 I
Main Strawberry	Tuolumne	S. Fk. Stanislaus R.	18,600 P
Mammoth Pool	Fresno	San Joaquin River	123,000 IP
Mariposa	Mariposa	Mariposa Creek	15,000 F
Mathews	Riverside	Cajalco Creek	182,000 M
McCloud	Shasta	McCloud River	35,300 P
Melones	Calaveras	Stanislaus River	112,500 IP
Monticello	Napa	Putah Creek	1,600,000 I
Morena	San Diego	Cottonwood Creek	50,206 M
Morris	Los Angeles	San Gabriel River	35,171 DM
Nacimiento	S. Luis Obispo	Salinas River	350,000 DIR
New Hogan	Calaveras	Calaveras River	320,000 IR
North Fork	Placer	N. Fk. American R.	14,600 F
Ole Indian	Lassen	Hamilton Creek	24,800 P
Oroville	Butte	Feather River	3,500,000 FIP
O'Shaughnessy	Tuolumne	Tuolumne River	360,000 DPM
Owen	Stanislaus	Tuolumne River	49,000 I
Palo Verde	Riverside	Colorado	-- --
Pardee	Amador	Mokelumne River	210,000 M
Parker	San Bernard.	Colorado River	717,000 PM
Peters	Marin	Lagunitas Creek	16,500 M
Pine Flat	Fresno	Kings River	1,000,000 F
Pit #3	Shasta	Pit River	40,600 P
Pit #7	Shasta	Pit River	34,000 P
Prado	Riverside	Santa Ana River	223,000 F
Puddingstone	Los Angeles	Walnut Creek	17,190 F
Railroad Canyon	Riverside	San Jacinto River	14,000 DI
Relief	Tuolumne	Relief Creek	15,122 P
Rollins	Nevada	Bear River	66,000 IP
Saddlebag	Mono	Leevining Creek	11,138 IP
Salt Springs	Amador	N. Fk. Mokelumne R.	139,400 P
Salt Springs Val.	Calaveras	Rock Creek	10,900 IM
Salinas	S. Luis Obispo	Salinas River	26,000 M
San Andreas	San Mateo	San Andreas Creek	18,500 DM
San Gabriel #1	Los Angeles	San Gabriel River	43,825 F
San Pablo	Contra Costa	San Pablo Creek	43,193 DM
Santa Fe	Los Angeles	San Gabriel River	33,000 F
Santa Felicia	Ventura	Piru Creek	100,000 DIP
Santiago Creek	Orange	Santiago Creek	25,000 I
San Vicente	San Diego	San Vicente Creek	90,234 DM
Savage	San Diego	Otay River	56,326 DM
Scott	Lake	Eel River	93,724 P
Scotts Flat	Nevada	Deer Creek	27,400 DIP
Sepulveda	Los Angeles	Los Angeles River	17,400 F
Shasta River	Siskiyou	Shasta River	72,000 I
Shaver Lake	Fresno	Stevenson Creek	135,283 P
Shasta Reservoir	Shasta	Sacramento River	4,492,700 IP
Silver Lake	Amador	Silver Fork River	11,800 P
Sly Creek	Butte	Lost Creek	65,050 IMP
Sly Park	El Dorado	Sly Park Creek	41,033 F
Stony Gorge	Glenn	Stony Creek	50,200 I
Sutherland	San Diego	Santa Ysabel Creek	29,000 M
Sweetwater, Main	San Diego	Sweetwater River	27,689 DIM
Terminus	Tulare	Kaweah River	150,000 FIR
Tinemaha	Inyo	Owens River	16,405 DM
Trinity	Trinity	Trinity River	2,500,000 IMD
Tule Lake	Lassen	Cedar Creek	39,500 IR
Twitchell	S. Barbara	Cuyama River	250,000 IMD
Tulloch	Calaveras	Stanislaus River	68,400 IP
Twin Lake	Alpine	Silver Fork	21,581 P
Union Valley	El Dorado	Silver Creek	271,000 MP
Upr. Crystal Spgs.	San Mateo	Lagune Creek	15,500 DM
Upr. San Leandro	Alameda	San Leandro Creek	41,436 DM
Vermillion	Fresno	Mono Creek	125,000 DP
West Valley	Modoc	West Valley Creek	17,700 I
Whale Rock	S. Luis Obispo	Old Creek	40,000 DIM
Whiskeytown	Trinity	Clear Creek	250,000 IMD
Whittier Narrows	Los Angeles	San Gabriel	35,000 F
Wishon	Fresno	N. Fk. Kings River	128,000 P
Woodward	Stanislaus	Simmons Creek	35,000 I

*USE:

D - Domestic
F - Flood Control
I - Irrigation
In - Industrial
M - Municipal
Mi - Mining
P - Power
R - Recreation

SOURCE: California Department of Water Resources, Dams Within Jurisdiction of the State of California, Bul. No. 17, Sacramento: January 1, 1958, 59 pp.

California Department of Water Resources, Major Dams and Reservoirs of California, Sacramento: September, 1959, 4pp.